Topical Research Meeting: Prospects in Neutrino Physics

(NuPhys2013)

Journal of Physics: Conference Series Volume 598

London, United Kingdom
19 – 20 December 2013

ISBN: 978-1-5108-0255-1
ISSN: 1742-6588

Printed from e-media with permission by:

Curran Associates, Inc.
57 Morehouse Lane
Red Hook, NY 12571

Some format issues inherent in the e-media version may also appear in this print version.

Printed by Curran Associates, Inc. (2015)

For permission requests, please contact the Institute of Physics
at the address below.

Institute of Physics
Dirac House, Temple Back
Bristol BS1 6BE UK

Phone: 44 1 17 929 7481
Fax: 44 1 17 920 0979

techtracking@iop.org

Additional copies of this publication are available from:

Curran Associates, Inc.
57 Morehouse Lane
Red Hook, NY 12571 USA
Phone: 845-758-0400
Fax: 845-758-2634
Email: curran@proceedings.com
Web: www.proceedings.com

TABLE OF CONTENTS

KEYNOTE TALKS

POSTER CONTRIBUTIONS

Journal of Physics: Conference Series **598** (2015) 011001 doi:10.1088/1742-6596/598/1/011001

Prospects in Neutrino Physics 2013 – NuPhys2013

19-20 December 2013, Institute of Physics, London, UK

The first "Prospects in Neutrino Physics 2013 – NuPhys2013" conference was held at the Institute of Physics, IoP, London, 19-20 December 2013 and was attended by about 130 delegates from institutions worldwide. Lunch and coffee breaks allowed discussions among delegates and speakers to take place in an informal setting.

This conference is unique in discussing the worldwide strategy to address unresolved issues in neutrino physics, and shape the future directions of particle physics. We discussed the current status and focussed especially on the prospects of future experiments, their performance and physics reach. It is particularly timely due to the recent measurements in neutrino physics and planned worldwide experiments.

The following topics were addressed:

- Theory and Phenomenology Perspectives
- Future Long and Short Baseline Neutrino Oscillation Experiments
- Reactor neutrino and flux
- Neutrinoless double beta decays
- Solar, atmospheric, supernova neutrinos
- Neutrino cosmology

in which both the phenomenological and experimental aspects were equally addressed.

World-leading experts in the different neutrino areas were invited to give review talks. To encourage and facilitate the participation of early-career researchers and PhD students, a poster session formed a key aspect of this meeting.

The conference was organized by Francesca Di Lodovico and Silvia Pascoli. It was sponsored by the IoP through their Topic Research Meeting Grant, and also supported by Durham IPPP, ERC-207282, FP7 invisibles project, Queen Mary University of London.

THE INSTITUTE OF PHYSICS

Peer review statement

All papers published in this volume of *Journal of Physics: Conference Series* have been peer reviewed through processes administered by the proceedings Editors. Reviews were conducted by expert referees to the professional and scientific standards expected of a proceedings journal published by IOP Publishing.

Topical Research Meeting: Prospects in Neutrino Physics (NuPhys2013) IOP Publishing
Journal of Physics: Conference Series **598** (2015) 012001 doi:10.1088/1742-6596/598/1/012001

Theory Perspective on Flavour

Ferruccio Feruglio

Dipartimento di Fisica e Astronomia 'G. Galilei', Università di Padova, Via Marzolo 8, I-35131 Padua, Italy

INFN, Sezione di Padova, Via Marzolo 8, I-35131 Padua, Italy

E-mail: `feruglio@pd.infn.it`

Abstract. An overview of the flavour problem is presented, with emphasis on the theoretical efforts to find a satisfactory description of fermion masses and mixing angles.

1. Introduction

The flavour problem exhibits two aspects. The first one is the origin of the parameters in the flavour sector of the SM, minimally extended to include massive Majorana neutrinos. Out of the 22 independent parameters, which will be collectively called $\mathcal{Y}_i$ or Yukawa couplings, 18 have already been measured. Only four parameters, the absolute scale of neutrino masses, the Dirac CP-violating phase and, possibly, two Majorana phases are still unknown. We are looking for a more economic description, perhaps related to a new principle, such as the gauge principle. Gauge invariance and renormalizability allow to describe strong and electroweak interactions of fifteen different fermion species in terms of only three parameters, augmented to five after including the electroweak symmetry breaking sector. Nothing similar exists so far in the flavour sector. The second aspect of the flavour problem is related to the new particle threshold around the TeV scale predicted by all SM extensions addressing the hierarchy problem. Once new TeV particles transforming non-trivially in flavour space are introduced, it is very difficult to maintain the almost perfect agreement between predictions and observations that reigns in the SM. New sources of FCNC and CP violations appear and the task is to keep them at an acceptable level. In this talk the focus will be on the first aspect, I will comment only shortly on the second one. Also, I'm not aiming at reviewing all the existing models, but rather in revisiting the main ideas, as an introduction to more specialized presentations at this workshop.

There are different kinds of approach to the flavour problem and here we can mention two of them. The first one is reductionist: the Yukawa couplings $\mathcal{Y}_i$ should be deduced from first principles. We postulate the existence of a fundamental theory from which $\mathcal{Y}_i$ can be uniquely determined. Either by proceeding directly from the candidate theory or by appealing to some symmetry or dynamical principle $\mathcal{Y}_i$ can be computed in terms of a small set of input parameters. Probably the most striking fact about this program is that nothing approaching a standard theory of $\mathcal{Y}_i$ exists, despite decades of experimental progress and theoretical efforts.

In the second approach a major role is played by chance. There are many variants and practical implementations of this approach. The Yukawa couplings $\mathcal{Y}_i$ are typically mapped to a large number of order-one parameters that are considered as irreducible unknowns, like in models with Froggatt-Nielsen abelian flavour symmetries or with fermions living in extra dimensions.

Also the paradigm of partial compositeness falls into this class. By scanning the order-one parameters we get probability distributions for masses and mixing angles. Alternatively we start from a fundamental theory, like string theory, which possesses a vast landscape of solutions, with no privileged ground state. The observed Yukawa couplings become environmental quantities and cannot be predicted, like the relative sizes of the solar planetary orbits. We are allowed to ask much less ambitious questions. For instance, if we have knowledge of the statistical distribution of $\mathcal{Y}_i$ in the fundamental theory, we can ask how typical are the Yukawa couplings that we observe. Conversely, barring anthropic selections, we might assume that the observed $\mathcal{Y}_i$ are typical and try to deduce information on the statistical distribution of $\mathcal{Y}_i$ in the fundamental theory.

In this talk I will adopt the point of view that the Yukawa couplings can be constrained to some extent by a flavour symmetry. The largest possible flavour symmetry of a theory with the particle content of the SM is $G_{MFV} = \mathrm{U}(3)^5$ and corresponds to the unrealistic limit in which the Yukawa couplings are turned off. The observed fermion masses and mixing angles break G_{MFV} almost completely to a residual symmetry that includes at most the weak hypercharge and, maybe, the combination B-L. Hence, in any realistic model based on flavour symmetries, the flavour symmetry group $G_f \subseteq G_{MFV}$ is broken. In predictive models the breaking is spontaneous, and it occurs through the VEVs of a set of scalar fields φ transforming non-trivially under G_f. The VEVs $\langle\varphi\rangle$ are determined by minimizing a G_f-invariant energy functional $V(\varphi)$. The Yukawa couplings becomes dynamical variables evaluated at the minimum of $V(\varphi)$: $\mathcal{Y}(\langle\varphi\rangle/\Lambda_f)$, Λ_f representing the scale of flavour physics. A huge number of models can be constructed according to this set of rules, depending on the choice of G_f (global, local, continuous, discrete, abelian, non-abelian), and on the choice of representations for scalars and fermions. A recent attempt to start from the full G_{MFV} symmetry and to study the stationary points of a general G_{MFV}-invariant potential is described in ref. [1].

2. Lessons from the quark sector

An empirical evidence for G_f comes from the quark sector and was pointed out long ago by Froggatt and Nielsen [2]. In their pioneering work Froggatt and Nielsen observed that all the small dimensionless parameters of the quark sector such as the quark mass ratios and the CKM mixing angles can be interpreted as powers of a small symmetry breaking parameter $\lambda \approx 0.2$. A scalar field φ, carrying by convention a negative unit of an abelian charge, develops a VEV that can be parametrized as

$$\lambda = \langle\varphi\rangle/\Lambda_f < 1 \qquad\qquad FN(\varphi) = -1 \quad . \tag{1}$$

Quarks carry non-negative $\mathrm{U}(1)_{FN}$ charges (the case with charges of both signs can be discussed as well)

$$FN(X_i) \geq 0 \qquad\qquad (X_i = Q_i, U_i^c, D_i^c) \quad . \tag{2}$$

Under these assumptions the quark Yukawa couplings $y_{u,d}$ are given by:

$$y_u = F_{U^c} Y_u F_Q \quad , \qquad\qquad y_d = F_{D^c} Y_d F_Q \quad , \tag{3}$$

where $Y_{u,d}$ are complex matrices with entries of order one, undetermined by the $\mathrm{U}(1)_{FN}$ symmetry, while F_X are real diagonal matrices, completely specified in terms of λ by the charges $FN(X_i)$:

$$F_X = \left(\begin{array}{ccc} \lambda^{FN(X_1)} & 0 & 0 \\ 0 & \lambda^{FN(X_2)} & 0 \\ 0 & 0 & \lambda^{FN(X_3)} \end{array}\right) \qquad\qquad (X_i = Q_i, U_i^c, D_i^c) \quad . \tag{4}$$

Journal of Physics: Conference Series **598** (2015) 012001 doi:10.1088/1742-6596/598/1/012001

The small quark mass ratios and quark mixing angles originate from the hierarchical structure of the matrices F_X. Indeed, by taking $FN(Q_1) > FN(Q_2) > FN(Q_3) \geq 0$ we get

$$(V_{u,d})_{ij} \approx \frac{F_{Q_i}}{F_{Q_j}} < 1 \quad (i < j) \tag{5}$$

for the matrices $V_{u,d}$ defining the CKM mixing matrix $V_{CKM} = V_u^\dagger V_d$. Independently from the specific charge choice, this framework predicts

$$V_{ud} \approx V_{cs} \approx V_{tb} \approx O(1) \qquad\qquad V_{ub} \approx V_{td} \approx V_{us} \times V_{cb} \quad , \tag{6}$$

the last equality being correct within a factor of two. With $\lambda \approx 0.2$, the correct order of magnitudes of the V_{CKM} matrix elements can be reproduced by choosing, for instance, $FN(Q) = (3,2,0)$. The correct order of magnitudes of the quark and charged lepton mass ratios, up to a couple of moderate tunings, can be reproduced by choosing

$$FN(U^c) = FN(E^c) = FN(Q) = (3,2,0) \qquad\qquad FN(D^c) = FN(L) = (2,0,0) \quad . \tag{7}$$

Notice that this charge assignment is also compatible with an SU(5) grand unified theory. The construction relies on a spontaneously broken abelian flavour symmetry, but the final results (3-6) are valid in a more general context, where no symmetry is present to start with. For example, if we consider split fermions in a warped extra dimension (a flat metric for the extra dimension works equally well), we can reproduce the same pattern of Yukawa coupling of eq. (3) with matrices F_X now given by

$$F_{X_i} = \sqrt{\frac{1-2c_i}{1-(R/R')^{1-2c_i}}} \approx \begin{cases} O(1) & c_i < 1/2 \\ (R/R')^{c_i-1/2} \ll 1 & c_i > 1/2 \end{cases} \tag{8}$$

where c_i are the bulk quark masses in units of $1/R$ and R, R' are length scales of the order of the inverse Planck mass and the inverse TeV scale, respectively. The Higgs field is localized on the TeV brane and has generic $O(1)$ Yukawa couplings $Y_{u,d}$ with the bulk quark fields. The role of the Froggatt-Nielsen charges is here played by the bulk masses c_i that determine the profile along the extra dimension of the zero-mode wave functions. There is no flavour symmetry: the hierarchical structure of quark masses and mixing angles is dictated by geometry in the compact space. Similarly, in the partial compositeness scenario, light fermions get hierarchical masses from the mixing between an elementary sector and a composite one. The composite sector has unsuppressed $O(1)$ interactions with the Higgs field and the Yukawa couplings of the light fermions have the structure given in eq. (3) with matrices F_X parametrizing the elementary-composite mixing. Here again symmetry considerations are absent.

In the previous examples the anarchical pattern of $Y_{u,d}$ results in strong bounds on the scale of new physics Λ_{NP} associated to particles carrying flavour quantum numbers and representing new sources of FCNC and/or CP violation. When the quark Yukawa couplings are described by eq. (3), the maximal flavour symmetry felt by quarks can be as large as $\mathrm{U}(3)^3$, the relevant subgroup of G_{MFV}. However there are more spurions than in MFV, the irreducible ones including now at least F_Q, F_{U^c}, F_{D^c}, Y_u and Y_d. One of the most dangerous effects originates from the effective operator

$$\frac{1}{\Lambda_{NP}^2}(\overline{Q}F_Q^\dagger\gamma_\mu F_Q Q)(\overline{D^c}F_{D^c}^\dagger\gamma^\mu F_{D^c}D^c) \approx \frac{1}{\Lambda_{NP}^2\langle Y_d^2\rangle}\frac{2m_d m_s}{v^2}(\overline{s_L}d_R)(\overline{s_R}d_L) + ... \tag{9}$$

$\langle Y_d^2\rangle$ representing an average $O(1)$ coupling. The contribution of this operator to the CP-violating ϵ_K parameter is enhanced at the level of both the hadronic matrix element and the QCD corrections. Assuming a generic $O(1)$ phase for the overall coefficient we need

$$\langle Y_d\rangle\, \Lambda_{NP} > 20 \text{ TeV} \tag{10}$$

not to spoil the SM prediction for ϵ_K. This, together with other constraints, suggests that a fully anarchical pattern in $Y_{u,d}$ is probably not tenable if new flavoured physics at the TeV scale is present.

In summary, the pattern of masses and mixing angles in the quark sector is well described by the map in eq. (3), in terms of input parameters of order one, the matrix elements of $Y_{u,d}$. Such a map can be realized in several different frameworks and does not necessarily needs an underlying symmetry. The setup is compatible with grand unification and with the known solution to the gauge hierarchy problem. On the weak side, additional ingredients are probably needed to control the new sources of FCNC and CP-violations arising from new physics at the TeV scale. Moreover all entries of $Y_{u,d}$ are independent free parameters and it is not possible to make absolute predictions, beyond the order-of-magnitude accuracy. This is clearly a major limitation, since we would like to test the theory at the level of the best available experimental precision.

3. From quarks to leptons

In the lepton sector we have no evidence for strong hierarchies in mixing angles or in neutrino masses. Hierarchy shows up at the level of charged lepton masses. In terms of the suppression factors F_{X_i} this means

$$F_{E_1^c} > F_{E_2^c} > F_{E_3^c} \geq 0 \qquad \text{and} \qquad F_{L_1} \approx F_{L_2} \approx F_{L_3} \quad . \tag{11}$$

Table 1. Possible choices of FN charges for the $\bar{5}$ representation in a class of SU(5) grand unified models, from ref. [4]. The second column shows the value of the FN symmetry breaking parameter optimizing the fit to fermion masses and mixing angles.

$FN(\bar{5})$	λ
(0,0,0)	-
(1,0,0)	0.25
(2,0,0)	0.35
(2,1,0)	0.45

Here we focus on Majorana neutrinos. A drastic realization of this picture is the framework of Anarchy [3], which in this language can be defined by

$$F_{L_1} = F_{L_2} = F_{L_3} \quad . \tag{12}$$

As a consequence the mass matrix for light neutrinos is

$$m_\nu = \begin{pmatrix} O(1) & O(1) & O(1) \\ O(1) & O(1) & O(1) \\ O(1) & O(1) & O(1) \end{pmatrix} m_0 \quad , \tag{13}$$

with undetermined order-one matrix elements. This implies mixing angles and neutrino mass ratios of $O(1)$, in rough agreement with the data. No special values for these quantities is expected and in this respect the non-vanishing, sizable value of θ_{13} and the departure of θ_{23} from $\pi/4$ exhibited by the present data have strengthened the case for hierarchy. Anarchy represents an extreme possibility and milder realization of the relations (11) are possible. For instance, in the context of SU(5) grand unified models, with a Froggatt-Nielsen $U(1)_{FN}$ abelian symmetry, neutrino masses and mixing angles can be reproduced, at the level of order of magnitudes, by several choices of the FN charges for the $\bar{5}$ multiplets hosting the lepton doublets, as shown in table 1. FN charges for fermions in the 10 representations can be suitably chosen so that, by varying the unknown order-one parameters, reasonable distributions for charged lepton mass ratios, quark mass ratios and quark mixing angles are also obtained. A naive comparison of the distributions for neutrino masses and mixing angles with data do not appear to favor anarchy over the other possible charge assignments.

If this framework also comprises new flavoured particles at the TeV scale, severe bounds from LFV apply. The irreducible sources of flavour violation in the lepton sector include the matrices Y_e, F_{E^c} and F_L and LFV can occur even in the limit of vanishing neutrino masses. Notice that we have no LFV in MFV in this limit, since in MFV the only available spurion in the lepton sector is Y_e. The dipole operator contributing to LFV is

$$\frac{e}{\Lambda_{NP}^2} E^c \sigma_{\mu\nu} F^{\mu\nu} (F_{E^c} Y_e Y_e^\dagger Y_e F_L) H^\dagger L \quad . \tag{14}$$

The charged lepton mass matrix is proportional to $(F_{E^c} Y_e F_L)$. In general the combinations $(F_{E^c} Y_e F_L)$ and $(F_{E^c} Y_e Y_e^\dagger Y_e F_L)$ are not diagonal in the same basis, not even in the case of universal F_L in eq. (12), and radiative decays of muon and tau are expected. Agreement with the most constraining upper bound, $BR(\mu \to e\gamma) < 5.7 \times 10^{-13}$, requires Λ_{NP} well above 10 TeV.

Table 2. Results of global fits to neutrino oscillation data from ref. [5] and ref. [6], for normal [NO] and inverted [IO] ordering.

	[5]	[6]
$\sin^2\theta_{13}$	$0.0231^{+0.0019}_{-0.0019}$	$0.0234^{+0.0022}_{-0.0018}$ [NO] $0.0239^{+0.0021}_{-0.0021}$ [IO]
$\sin^2\theta_{23}$	$0.446^{+0.008}_{-0.008} \oplus 0.593^{+0.027}_{-0.043}$	$0.437^{+0.033}_{-0.023}$ [NO] $0.455^{+0.139}_{-0.031}$ [IO]
δ_{CP}/π	$1.48^{+0.31}_{-0.35}$	$1.39^{+0.38}_{-0.27}$ [NO] $1.31^{+0.29}_{-0.33}$ [IO]

As in the quark sector, a completely anarchical matrix Y_e and flavoured physics at the TeV scale are difficult to reconcile. A sufficient condition for the absence of LFV is that Y_e, F_{E^c} and F_L are diagonal in the same basis, as suggested in some models. Alternatively we can look for special forms of these matrices dictated by some symmetry requirements.

4. Discrete symmetries

The data from neutrino oscillations before 2012 were supporting flavour symmetries, especially through the indication of a vanishing reactor angle θ_{13} and a maximal atmospheric mixing angle θ_{23}, features that are difficult to attribute to an underlying theory based on pure chance. Today we know with accuracy that θ_{13} is neither vanishing nor particularly small, its size being comparable to that of the Cabibbo angle. Global fits show that the atmospheric mixing angle is probably deviating from the maximal value by few degrees, see table 2. While CP violating effects in the lepton sector are still undetected, recent global fits hint to a maximal Dirac CP-violating phase δ_{CP}, though the whole range from 0 to 2π is still allowed at 3σ. Before the measurement of θ_{13} a particularly attractive lepton mixing pattern was the tribimaximal one

$$U_{TB} = \begin{pmatrix} \frac{2}{\sqrt{6}} & \frac{1}{\sqrt{3}} & 0 \\ -\frac{1}{\sqrt{6}} & \frac{1}{\sqrt{3}} & -\frac{1}{\sqrt{2}} \\ -\frac{1}{\sqrt{6}} & \frac{1}{\sqrt{3}} & \frac{1}{\sqrt{2}} \end{pmatrix} \approx \begin{pmatrix} 0.82 & 0.58 & 0 \\ -0.41 & 0.58 & -0.71 \\ -0.41 & 0.58 & 0.71 \end{pmatrix} \quad . \tag{15}$$

The good agreement between TB mixing and pre-2012 data strongly supported the idea that the true mixing matrix could be described in terms of small corrections to a LO mixing matrix U^0_{PMNS}, which could be derived from symmetry considerations. The simplest way to reproduce the TB mixing pattern is by exploiting discrete flavour symmetries [7, 8]. The theory is invariant under a discrete flavour symmetry G_f, broken down in such a way that neutrino and charged lepton sectors have different residual symmetries G_ν and G_e, at least in a LO approximation where small effects are neglected. If neutrinos are of Majorana type, the most general group leaving m_ν invariant and the individual masses m_i unconstrained is $Z_2 \times Z_2$, a finite group. The subgroup G_e can be continuous, but G_e discrete is the simplest option. We require a sufficiently large G_e to distinguish the three charged leptons. For instance we can choose $G_e = Z_n$ $(n \geq 3)$ or $G_e = Z_2 \times Z_2$. Once G_e and G_ν have been chosen inside G_f, the embedding automatically fixes the relative alignment of $m_l^\dagger m_l$ and m_ν in flavour space. Lepton masses are unconstrained and U^0_{PMNS} is determined up to Majorana phases and up to permutations of rows and columns. This freedom apart, we can predict the three mixing angles θ^0_{ij} and the Dirac phase δ^0_{CP}. In most concrete models, where symmetry breaking is achieved via VEVs of a set of flavons, the LO results are modified by small corrections

$$U_{PMNS} = U^0_{PMNS} + O(u) \quad . \tag{16}$$

Before 2012, in the specific case $U^0_{PMNS} = U_{TB}$ these corrections were expected to be very small, of the order of few percent, not to spoil the good agreement in the predicted value of the solar mixing angle. On this basis the simplest models reproducing U_{TB} at the LO predicted θ_{13} not larger than few degrees, now proven to be wrong by experiments. Discrete flavour symmetries can also be extended to quarks and even incorporated in GUTs, but in the existing constructions the symmetry has to be badly broken in the quark sector. Discrete flavour symmetries are also relevant in the so called indirect models [8]. In this case the breaking of G_f leaves no residual symmetries and the role of the flavour group is mainly to get specific vacuum alignment of the scalar fields that control fermion masses.

Several modifications of the simplest models based on discrete symmetries have been proposed to match the most recent data. If we keep adopting $U^0_{PMNS} = U_{TB}$ as LO approximation, perhaps the most economic way to reproduce the actual value of θ_{13} is to introduce large correction terms, $O(u) \approx 0.2$. This is also viable in some scheme where U^0_{PMNS} differs substantially from U_{TB}, such as the so-called bimaximal mixing. Introducing large corrections has the disadvantage that beyond the LO the number of independent contributions is generally quite large. If their typical size is about 0.2, all mixing angles tend to be affected by generic corrections of this type and predictability is lost. Moreover large correction terms are dangerous if new sources of flavour changing and/or CP violation are present at the TeV scale.

Another possibility is to look for alternative LO approximations where θ_{13} is closer to the measured value. Remarkably, several groups G_f giving rise to LO approximations closer to the data have been found. Of particular interest is a special form of trimaximal (TM) mixing:

$$U^0_{PMNS} = U_{TB}\, U_{13}(\alpha,\delta) = U_{TB} \begin{pmatrix} \cos\alpha & 0 & e^{i\delta}\sin\delta \\ 0 & 1 & 0 \\ -e^{-i\delta}\sin\alpha & 0 & \cos\alpha \end{pmatrix} \quad , \tag{17}$$

reproduced by groups of the series $\Delta(6n^2)$ [9, 10, 11, 12]. For $n = (4, 8, 10)$ we have $\alpha = (\pm 1/12, \pm 1/24, \pm\pi/15)$ and $\sin^2\theta^0_{13} = (0.045, 0.011, 0.029)$. The Dirac phase is zero (modulo π).

A further possibility is to relax the symmetry requirements. S_4 is the smallest group reproducing TB mixing through the breaking down to $G_e = Z_3$ and $G_\nu = Z_2 \times Z_2$, whose generators are T and (S, U), respectively. In the basis where T and the charged leptons are diagonal, the element U coincides with the so-called $\mu\tau$ exchange symmetry, directly responsible for the vanishing of θ_{13} and for θ_{23} being maximal. We can avoid having $\theta_{13} = 0$ and $\theta_{23} = \pi/4$ if G_ν is a single Z_2 subgroup generated either by the element S or by the element SU. When the preserved parity is S, the mixing pattern, TM_2, is trimaximal and corresponds to that in eq. (17), with both α and δ unconstrained. When the preserved parity is SU, the mixing pattern, TM_1, is also of trimaximal type and is given by $U^0_{PMNS} = U_{TB}\, U_{23}(\alpha,\delta)$, where $U_{23}(\alpha,\delta)$ is the transformation analogous to $U_{13}(\alpha,\delta)$, acting in the 23 plane. The mixing angles and the Dirac phase are predicted in terms of (α,δ) and we get two relations among physical quantities, shown in Table 3. Explicit models based on A_4 realizing TM_2 breaking pattern were indeed proposed

Table 3. Sum rules for $TM_{1,2}$ mixing patterns.

TM_1	TM_2
$\sin^2\theta_{12} = \frac{1}{3} - \frac{2}{3}\sin^2\theta_{13} + O(\sin^4\theta_{13})$	$\sin^2\theta_{12} = \frac{1}{3} + \frac{1}{3}\sin^2\theta_{13} + O(\sin^4\theta_{13})$
$\sin^2\theta_{23} = \frac{1}{2} - \sqrt{2}\sin\theta_{13}\cos\delta_{CP} + O(\sin^2\theta_{13})$	$\sin^2\theta_{23} = \frac{1}{2} + \frac{1}{\sqrt{2}}\sin\theta_{13}\cos\delta_{CP} + O(\sin^2\theta_{13})$

before the measurement of θ_{13} [13]. The possibility of reducing the residual symmetry G_ν to Z_2 can be systematically investigated [14].

A more recent development consists in combining discrete and CP symmetries and exploring the symmetry breaking patterns such a combination can give rise to. A well-known example is that of the so-called $\mu\tau$ reflection symmetry [15, 16] (not to be confused with the $\mu\tau$ exchange symmetry), which exchanges a muon (tau) neutrino with a tau (muon) antineutrino in the charged lepton mass basis. If such a symmetry is imposed, the atmospheric mixing angle is predicted to be maximal, while θ_{13} is in general non-vanishing for a maximal Dirac phase δ and the Majorana phases vanish. The solar mixing angle remains unconstrained.

A general formalism which combines CP and flavour symmetries, proposed in ref. [17], can be used to constrain the lepton mixing matrix. A theory symmetric under CP and under a discrete flavour group G_f is assumed to have residual symmetries G_e, generated by some elements Q_i and $G_\nu = Z_2 \times CP$, generated by a parity Z and a CP transformation X. The action of X in flavour space can be non-trivial and should respect a set of consistency conditions [18, 19]. The residual symmetries G_e and G_ν imply the following conditions on $m_l^\dagger m_l$ and m_ν:

$$Q_i^\dagger (m_l^\dagger m_l) Q_i = (m_l^\dagger m_l) \quad , \qquad Z^T m_\nu Z = m_\nu \quad , \qquad X m_\nu X = m_\nu^* \quad . \tag{18}$$

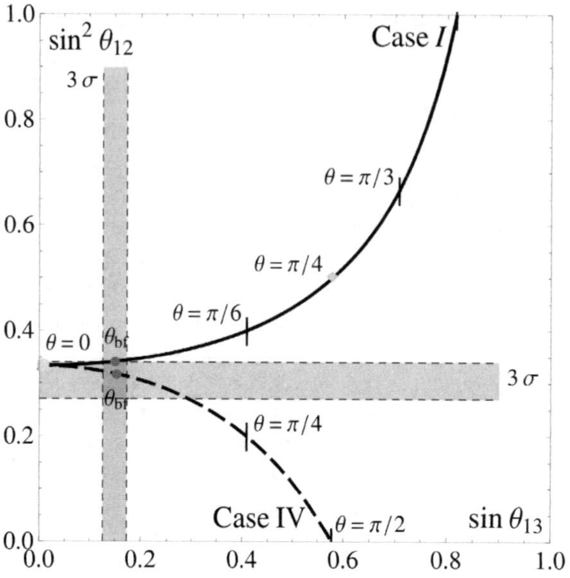

Figure 1. Results for the mixing parameters $\sin\theta_{13}$, $\sin^2\theta_{12}$ for Case I (straight line) and Case IV (dashed line). We mark the value $\theta_{\rm bf}$ of the parameter θ for which the χ^2 functions have a global minimum with a red dot. $3\,\sigma$ ranges for the mixing angles are also shown

These conditions are strong enough to completely determine U^0_{PMNS}, up to one real parameter θ, ranging from 0 to π:

$$U^0_{PMNS} = U^0_{PMNS}(Q_i, Z, X, \theta) \qquad\qquad 0 \leq \theta \leq \pi \quad . \tag{19}$$

Mixing angles and phases, both Dirac and Majorana, are then predicted as a function of θ, modulo the ambiguity related to the freedom of permuting rows and columns and to the intrinsic parity of neutrinos. The formalism is completely invariant under any change of basis in field space. The physical results only depend on the initial symmetry and the residual symmetries specified by (Q_i, Z, X). An interesting example is provided by $G_f = S_4$. An exhaustive analysis has been presented in ref. [17]. The residual symmetries can be chosen as $G_e = Z_3$, generated by the element T, and $G_\nu = Z_2 \times CP$, generated by (Z, X). The parity transformation Z can be either S (case I) or SU (case IV) and a consistent CP transformation X acting on the lepton doublets coincides with the $\mu - \tau$ reflection symmetry in the basis where T (and the combination $m_l^\dagger m_l$) is diagonal. Thus the predicted mixing pattern has a maximal atmospheric mixing angle, a maximal Dirac phase, vanishing Majorana phases and there is a relation between the solar angle and the reactor angle, shown in fig. 1.

Conclusion

We have recently witnessed a decisive experimental progress in neutrino physics. The reactor angle is now precisely known and it is away from zero by many standard deviations. We also have a first indication favoring a non-maximal atmospheric mixing angle. For the first time global fits hint to a non-trivial Dirac phase. While these steps have been effective in ruling out many models of fermion masses and mixing angles, it is fair to say that no compelling and unique theoretical picture has emerged so far. Present data can still be described within widely different frameworks. There are several models where fermion masses and mixing angles are mapped into a large number of irreducible and unconstrained order-one parameters, thus explaining the observed hierarchies. This class of models has the advantage of providing a common description for both quarks and leptons, which can be easily incorporated in a grand unified theory. The weak points are the impossibility to test this idea beyond the order-of-magnitude level, and the probable incompatibility with new flavoured physics at the TeV scale. As an alternative possibility, we have models where the flavour sector becomes highly symmetric at a high scale, and a small number of relevant parameters provides a complete description. Within this class, models based on discrete symmetries, tailored to reproduce the features of the early data from neutrino oscillations, are less supported by data now. Several modifications of the simplest schemes to accommodate the present data are still possible and have the advantage of being quantitatively testable. The weak point of this approach is that it is probably too much centered around the neutrino mixing properties while no clear indication for the need of a discrete symmetry seems to come from the quark sector or from the lepton masses. Flavour remains a fascinating mystery, still resisting all our attempts to find the rationale underlying our observations, and certainly worth to be investigated from all possible angles.

Acknowledgments
Many thanks to the organizers of NuPhys2013 for providing an extremely nice environment in London. I would like to thank warmly my collaborators Guido Altarelli, Claudia Hagedorn, Isabella Masina, Luca Merlo, Robert Ziegler for the pleasant collaboration on which the material of this talk is based.

References

[1] R. Alonso, M. B. Gavela, G. Isidori and L. Maiani, JHEP **1311**, 187 (2013) [arXiv:1306.5927 [hep-ph]].
[2] C. D. Froggatt and H. B. Nielsen, Nucl. Phys. B **147**, 277 (1979).
[3] L. J. Hall, H. Murayama and N. Weiner, Phys. Rev. Lett. **84**, 2572 (2000) [hep-ph/9911341].
[4] G. Altarelli, F. Feruglio, I. Masina and L. Merlo, JHEP **1211**, 139 (2012) [arXiv:1207.0587 [hep-ph]].
[5] F. Capozzi, G. L. Fogli, E. Lisi, A. Marrone, D. Montanino and A. Palazzo, arXiv:1312.2878 [hep-ph].
[6] M. C. Gonzalez-Garcia, M. Maltoni, J. Salvado and T. Schwetz, JHEP **1212**, 123 (2012) [arXiv:1209.3023 [hep-ph]].
[7] G. Altarelli and F. Feruglio, Rev. Mod. Phys. **82**, 2701 (2010) [arXiv:1002.0211 [hep-ph]].
[8] S. F. King and C. Luhn, Rept. Prog. Phys. **76**, 056201 (2013) [arXiv:1301.1340 [hep-ph]].
[9] R. d. A. Toorop, F. Feruglio and C. Hagedorn, Phys. Lett. B **703**, 447 (2011) [arXiv:1107.3486 [hep-ph]].
[10] R. de Adelhart Toorop, F. Feruglio and C. Hagedorn, Nucl. Phys. B **858**, 437 (2012) [arXiv:1112.1340 [hep-ph]].
[11] M. Holthausen, K. S. Lim and M. Lindner, Phys. Lett. B **721**, 61 (2013) [arXiv:1212.2411 [hep-ph]].
[12] S. F. King, T. Neder and A. J. Stuart, Phys. Lett. B **726**, 312 (2013) [arXiv:1305.3200 [hep-ph]].
[13] Y. Lin, Nucl. Phys. B **824**, 95 (2010) [arXiv:0905.3534 [hep-ph]].
[14] D. Hernandez and A. Y. .Smirnov, Phys. Rev. D **86**, 053014 (2012) [arXiv:1204.0445 [hep-ph]].
[15] P. F. Harrison and W. G. Scott, Phys. Lett. B **535** (2002) 163 [hep-ph/0203209]; P. F. Harrison and W. G. Scott, Phys. Lett. B **547** (2002) 219 [hep-ph/0210197]; P. F. Harrison and W. G. Scott, Phys. Lett. B **594** (2004) 324 [hep-ph/0403278].
[16] W. Grimus and L. Lavoura, Phys. Lett. B **579** (2004) 113 [hep-ph/0305309].
[17] F. Feruglio, C. Hagedorn and R. Ziegler, JHEP **1307**, 027 (2013) [arXiv:1211.5560 [hep-ph]].
[18] G. Ecker, W. Grimus and H. Neufeld, J. Phys. A **20** (1987) L807; H. Neufeld, W. Grimus and G. Ecker, Int. J. Mod. Phys. A **3** (1988) 603.
[19] W. Grimus and M. N. Rebelo, Phys. Rept. **281** (1997) 239 [hep-ph/9506272].

Status of 3ν oscillation parameters at the end of 2013

F. Capozzi[1,2], G.L. Fogli[1,2], E. Lisi[2,*], A. Marrone[1,2], D. Montanino[3,4] and A. Palazzo[5]

[1]Dipartimento Interateneo di Fisica, Via Amendola 173, 70126 Bari, Italy
[2]Istituto Nazionale di Fisica Nucleare, Sezione di Bari, Via Orabona 4, 70126 Bari, Italy
[3]Dipartimento di Matematica e Fisica, Via Arnesano, 73100 Lecce, Italy
[4]Istituto Nazionale di Fisica Nucleare, Sezione di Lecce, Via Arnesano, 73100 Lecce, Italy
[5]Max-Planck-Institut für Physik, Föhringer Ring 6, 80805 München, Germany

*Speaker. Email: `eligio.lisi@ba.infn.it`

Abstract. Neutrino oscillation searches using a variety of sources (solar, atmospheric, accelerator and reactor neutrinos) have established a standard three-neutrino (3ν) mass-mixing framework and five of its parameters: the two squared mass gaps (δm^2, Δm^2) and the three mixing angles (θ_{12}, θ_{13}, θ_{23}). At present, a single class of experiments dominates each of these parameters, while only combined analyses of various (eventually all) data sets are needed to constrain the still unknown mass hierarchy [sign(Δm^2)], θ_{23} octant and CP-violating phase δ. We review the status of the known and unknown parameters — as emerging from a global analysis of the oscillation data available at the end of 2013 — and discuss the correlations and stability of the such parameters within different combinations of data sets.

1. Introduction

In just 15 years since the discovery of atmospheric neutrino oscillations, a new paradigm — the 3ν mass-mixing framework — has emerged in particle physics. Indeed, the vast majority of ν oscillation data can be explained by assuming that the three known flavor states $\nu_\alpha = (\nu_e, \nu_\mu, \nu_\tau)$ are mixed with three massive states $\nu_i = (\nu_1, \nu_3, \nu_3)$ via three mixing angles $(\theta_{12}, \theta_{13}, \theta_{23})$ and a possible CP-violating phase δ. The observed oscillations are driven by two independent differences between the squared masses m_i^2, which can be defined as $\delta m^2 = m_2^2 - m_1^2 > 0$ and $\Delta m^2 = m_3^2 - (m_1^2 + m_2^2)/2$, where $\Delta m^2 > 0$ and < 0 correspond to normal hierarchy (NH) and inverted hierarchy (IH), respectively.

At present (end of 2013), five of the above 3ν oscillation parameters have been measured, with an accuracy largely dominated by a specific class of experiments, namely: θ_{12} by solar data [1], θ_{13} by short-baseline (SBL) reactor data [2], θ_{23} by atmospheric data, mainly from Super-Kamiokande (SK) [3], δm^2 by long-baseline reactor data from KamLAND (KL) [2], and Δm^2 by long-baseline (LBL) accelerator data, mainly from MINOS [4] and T2K [5]. However, the mass hierarchy, the θ_{23} octant, and the CP-violating phase δ are still unknown.

In this context, global neutrino data analyses may be useful to assess the overall consistency and accuracy of the known parameters, as well as to squeeze possible hints about the unknown ones. In the following, we report and discuss the results of a recent global analysis which include all the data available at the time of this NuPhys Conference (December 2013) [6]. The reader is referred to [6] for further details and references not reported herein.

It should be noted that, in the 3ν framework, there are other unknowns not accessible to oscillation experiments, namely: the absolute neutrino mass scale [7, 8], the Dirac or Majorana nature of the neutrino fields [9, 10] and, in the latter case, the associated Majorana phases [9, 10, 11]. Current constraints on these unknowns, which are crucial for theoretical model building [10, 11, 12], will also be briefly commented below. Finally, it should be mentioned that some controversial results (not discussed herein) might indicate possible extensions of the above 3ν framework in terms of one or more additional mass states ν_j ($j \geq 4$), mostly sterile and with mass gaps at the (sub)eV scale. The reader is referred to [13] for an up-to-date discussion of the sterile neutrino phenomenology.

2. Global analysis: Methodology

In this Section we briefly discuss the various data sets and their combination in global fits.

LBL Acc. + Solar + KL data. The oscillation phenomenology of LBL accelerator experiments is dominated by the oscillation parameters (Δm^2, θ_{23}) in the $\nu_\mu \to \nu_\mu$ disappearance channel, supplemented by θ_{13} in the $\nu_\mu \to \nu_e$ appearance channel. However, the current accuracy of MINOS and T2K data requires that the oscillation probability is precisely calculated in terms of all the input parameters, including matter effects and subdominant terms driven by (δm^2, θ_{12}, δ). Since (δm^2, θ_{12}) are essentially fixed by the Solar and KL experiments, it makes sense to combine these data with LBL accelerator data from the very beginning. We remark that "Solar + KL" data provide a preference for $\sin^2\theta_{13} \sim 0.02$ in our analysis, which plays a role in the combination "LBL Acc. + Solar + KL," as discussed below.

Adding SBL reactor data. After the recent T2K observation of electron flavor appearance, the combination of LBL Acc. + Solar + KL data can provide a highly significant measurement of θ_{13} which, however, depends on the unknown CP violating phase δ and θ_{23} octant. SBL reactor experiments (Daya Bay, RENO, Double Chooz) provide (δ, θ_{23})-independent and accurate measurements of θ_{13}, which play a crucial role in the "LBL Acc. + Solar + KL + SBL Reac." combination.

Adding atmospheric neutrino data. Atmospheric data involve a very rich oscillation phenomenology in both appearance and disappearance modes involving ν_μ and ν_e. In principle, the high-statistics Super-Kamiokande experiment (phases I-IV) is thus sensitive to subleading effects related to the mass hierarchy, the θ_{23} octant and the CP phase δ. However, within the current experimental and theoretical systematic uncertainties, it remains difficult to disentangle and probe such small effects at a level exceeding $\sim 1\sigma$–2σ. Moreover, different and independent analyses of SK data, at comparable levels of refinement, do not necessarily provide similar hints about subleading effects. Therefore, we prefer to add these data only in the final "LBL Acc. + Solar + KL + SBL Reac. + SK Atm." combination, in order to separately gauge their effects on the various 3ν parameters.

Conventions for allowed regions. The data are compared to theoretical expectations via a refined χ^2 function which accounts for all known sources of correlated and uncorrelated uncertainties. In each of the above combined data analyses, the six oscillation parameters (Δm^2, δm^2, θ_{12}, θ_{13}, θ_{23}) are unconstrained in any given hierarchy (normal or inverted). Parameter ranges at N standard deviations are defined as $N\sigma = \sqrt{(}\chi^2 - \chi^2_{\rm min})$. This definition holds also in two-dimensional plots, where it is understood that the previous $N\sigma$ ranges are reproduced by projecting 2D contours over one parameter axis. All undisplayed parameters are marginalized away. Finally, the relative preference of the data for either NH or IH is measured by the quantity $\Delta\chi^2_{\rm I-N} = \chi^2_{\rm min}({\rm IH}) - \chi^2_{\rm min}({\rm NH})$, with the caveat that it cannot immediately be translated into "$N\sigma$" by taking the square root of its absolute value, because it refers to two discrete hypotheses [14].

3. Results on single oscillation parameters

In this Section we graphically report the results of our global analysis of increasingly richer data sets, grouped in accordance to the previous discussion, in terms of single oscillation parameters.

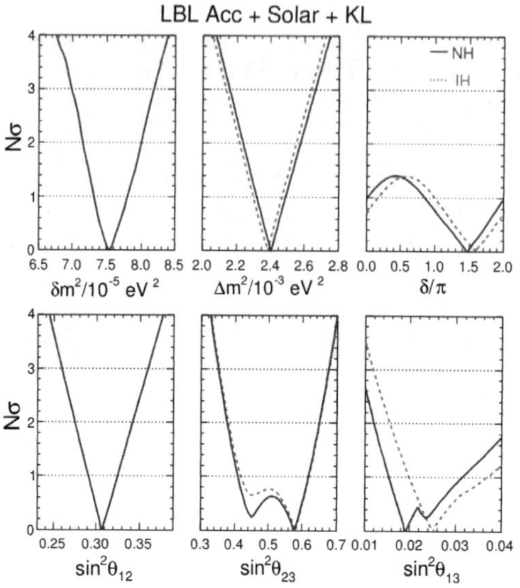

Figure 1. Combined 3ν analysis of LBL Acc. + Solar + KL data: Bounds on the oscillation parameters in terms of standard deviations $N\sigma$ from the best fit. Solid (dashed) lines refer to NH (IH). The horizontal dotted lines mark the 1σ, 2σ and 3σ levels for each parameter.

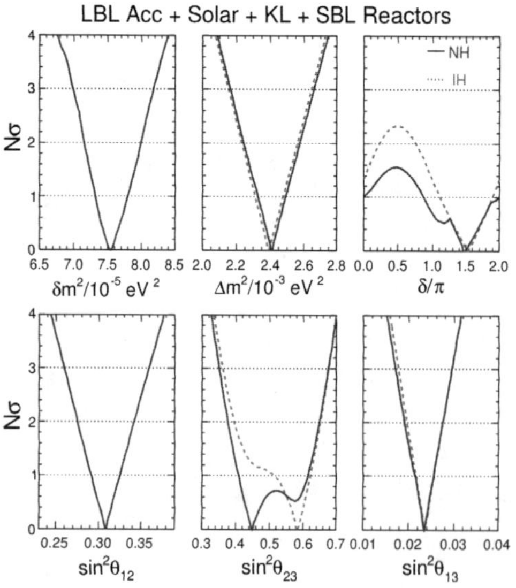

Figure 2. As in Fig. 1, but adding SBL reactor data.

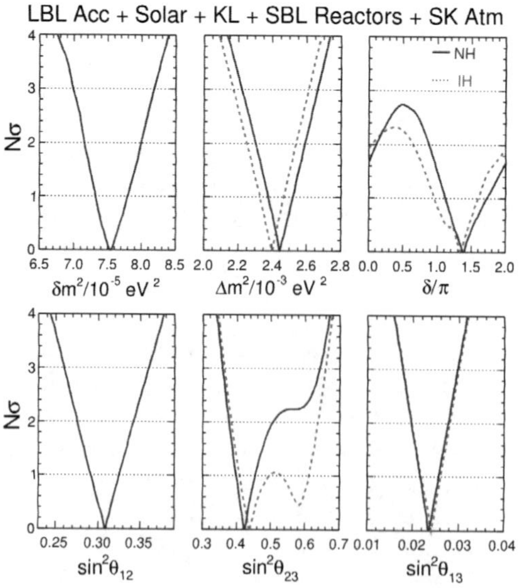

Figure 3. As in Fig. 2, but adding SK atmospheric data (global fit to all ν data).

Figures 1, 2 and 3 show the $N\sigma$ curves for the data sets defined in the previous section. In each figure, the solid (dashed) curves refer to NH (IH); the two curves basically coincide for the δm^2 and θ_{12} parameters, since they are determined by Solar+KL data which are largely insensitive to the hierarchy. For each parameter in Figs. 1–3, the more linear and symmetrical are the curves, the more gaussian is the associated probability distribution.

Figure 1 refers to the combination LBL Acc. + Solar + KL which, by itself, sets highly significant lower and upper bounds on all the oscillation parameters but δ. In this figure, the relatively strong appearance signal in T2K [5] dominates the lower bound on θ_{13}, and also drives the slight but intriguing preference for $\delta \simeq 1.5\pi$: indeed, for $\sin\delta \sim -1$, the CP-odd term in the $\nu_\mu \to \nu_e$ appearance probability is maximized. It should be noted that current MINOS appearance data generally prefer $\sin\delta > 0$ [4]; however, the stronger T2K appearance signal largely dominates in the global fit. On the other hand, MINOS disappearance data drive the slight preference for nonmaximal θ_{23}, as compared with nearly maximal θ_{23} in T2K [5]. The (even slighter) preference for the second θ_{23} octant is due to the interplay of LBL accelerator and Solar + KL data, as discussed in the next Section.

Figure 2 shows the results obtained by adding the SBL reactor data, which strongly reduce the θ_{13} uncertainty. Further effects of these data include: (*i*) a slightly more pronounced preference for $\delta \simeq 1.5\pi$ and $\sin\delta < 0$, and (*ii*) a swap of the preferred θ_{23} octant with the hierarchy ($\theta_{23} < \pi/4$ in NH and $\theta_{23} > \pi/4$ in IH). These features will be interpreted in terms of parameter covariances in the next Section.

Figure 3 shows the results obtained by adding the SK atmospheric data, thus obtaining the most complete data set. The main differences with respect to Fig. 2 include: (*i*) an even more pronounced preference for $\sin\delta < 0$, with a slightly lower best fit at $\delta \simeq 1.4\pi$; (*ii*) a slight reduction of the errors on Δm^2 and a relatively larger variation of its best-fit value with the hierarchy; (*iii*) a preference for θ_{23} in the first octant for both NH and IH, which is a persisting feature of our analyses. The effects (*ii*) and (*iii*) show that atmospheric neutrino data have the potential to probe subleading hierarchy effects, although they do not yet emerge in a stable or significant way. Table I summarizes in numerical form the results shown in Fig. 3.

In Figs. 1–3, an intriguing feature is the increasingly pronounced preference for nonzero CP violation with increasingly rich data sets, although the two CP-conserving cases ($\delta = 0, \pi$) remain allowed at $< 2\sigma$ in both NH and IH, even when all data are combined (see Fig. 3). It is worth noticing that the two maximally CP-violating cases ($\sin\delta = \pm 1$) have opposite likelihood: while the range around $\delta \sim 1.5\pi$ ($\sin\delta \sim -1$) is consistently preferred, small ranges around $\delta \sim 0.5\pi$ ($\sin\delta \sim +1$) appear to be disfavored (at $> 2\sigma$ in Fig. 3), In the next few years, the appearance channel in LBL accelerator experiments will provide crucial data to investigate these hints about ν CP violation [6], with relevant implications for models of leptogenesis [10].

From the comparison of Figs. 1–3 one can also notice a generic preference for nonmaximal mixing ($\theta_{23} \neq 0$), although it appears to be weaker than in our past analyses, essentially because the most recent T2K data prefer nearly maximal mixing, and thus "dilute" the opposite preference coming from MINOS and atmospheric data. Moreover, the indications about the octant appear to be somewhat unstable in different combinations of data. In the present analysis, only atmospheric data consistently prefer the first octant in both hierarchies, but the overall significance remains at the level $\sim 2\sigma$ in NH and is much lower in IH. These fluctuations show how difficult it is to reduce the allowed range of θ_{23}. In this context, the disappearance channel in LBL accelerator experiments will provide crucial data to address the issue of nonmaximal θ_{23} in the next few years [5, 15, 16].

Finally, we comment on the size of $\Delta\chi^2_{\mathrm{I-N}}$ which, by construction, is not apparent in Figs. 1–3. We find $\Delta\chi^2_{\mathrm{I-N}} = -1.3, -1.4, +0.3$, for the data sets in Figs. 1, 2, and 3, respectively. Unfortunately, such values are both small and with unstable sign, and do not provide us with any relevant indication about the hierarchy.

4. Selected parameter covariances

In this Section we show the allowed regions for selected couples of oscillation parameters, and discuss some interesting correlations.

Figure 4 shows the allowed regions in the plane $(\sin^2\theta_{23}, \sin^2\theta_{13})$. From left to right, the panels refer to increasingly rich data sets, while upper and lower panels refer to NH and IH, respectively. In the left panels, a slight negative correlation emerges from LBL appearance data, since the dominant oscillation amplitude contains a factor $\sin^2\theta_{23}\sin^2\theta_{13}$. The contours extend towards relatively large values of θ_{13}, especially in IH, in order to accommodate the relatively strong T2K appearance signal [5]. However, solar + KL data provide independent (although weaker) constraints on θ_{13} and, in particular, prefer $\sin^2\theta_{13} \sim 0.02$ in our analysis. This value is on the "low side" of the allowed regions and is thus responsible for the relatively high value of θ_{23} at best fit, namely, for the second-octant preference in both NH and IH. However, when current SBL reactor data are included in the middle panels, a slightly higher value of θ_{13} ($\sin^2\theta_{13} \simeq 0.023$) is preferred with very small uncertainties: this value is high enough to shift the best-fit of θ_{23} from the second to the first octant in NH, but not in IH. Finally, the inclusion of SK atmospheric data (right panels) provides in our analysis an overall preference for the first octant, which is however quite weak in IH. Unfortunately, as previously mentioned, the current hints about the θ_{23} octant do not appear to be particularly stable or convergent.

Figure 5 shows the allowed regions in the plane $(\sin^2\theta_{13}, \delta/\pi)$, which is at the focus of current research in neutrino physics. In the left panels there is a remarkable preference for $\delta \sim 1.5\pi$, where a compromise is reached between the relatively high θ_{13} values preferred by the T2K appearance signal, and the relatively low value preferred by solar + KL data. In the middle panel, SBL reactor data strengthen this trend by reducing the covariance between θ_{13} and δ. It is quite clear that we can still learn much from the combination of accelerator and reactor data in the next few years. Finally, the inclusion of SK atmospheric data in the right panels also adds some statistical significance to this trend, with a slight lowering of the best-fit value of δ.

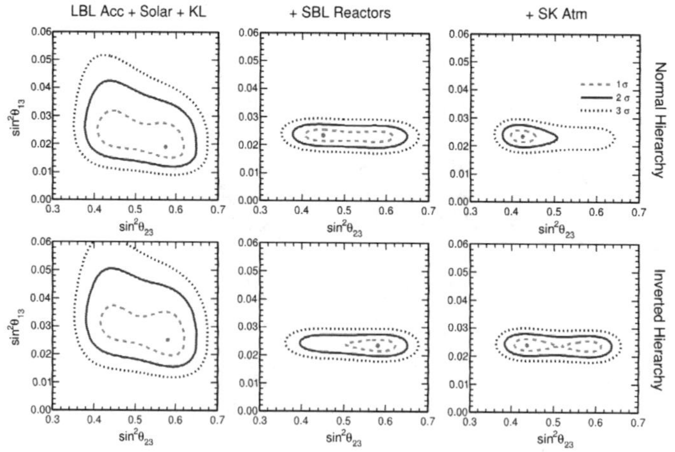

Figure 4. Results of the analysis in the plane charted by $(\sin^2\theta_{23}, \sin^2\theta_{13})$, all other parameters being marginalized away. From left to right, the regions allowed at 1, 2 and 3σ refer to increasingly rich datasets: LBL accelerator + solar + KamLAND data (left panels), plus SBL reactor data (middle panels), plus SK atmospheric data (right panels). Best fits are marked by dots. The three upper (lower) panels refer to normal (inverted) hierarchy.

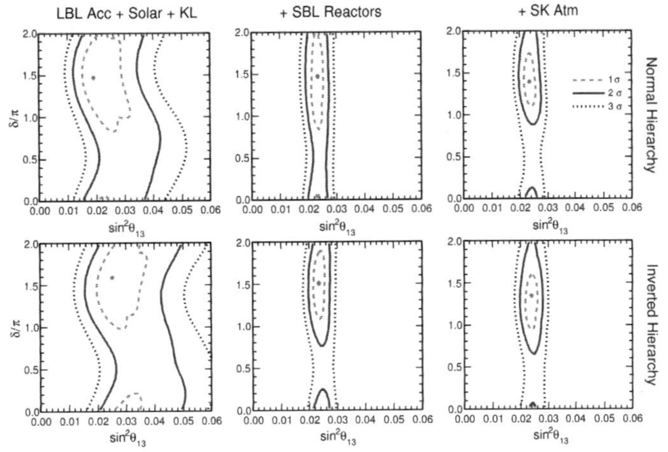

Figure 5. As in Fig. 4, but in the plane $(\sin^2\theta_{13},\ \delta/\pi)$.

5. Implications on absolute neutrino mass observables

In general, absolute neutrino masses can be probed via three main methods. The first, classical one is provided by β decay, sensitive to the so-called "effective electron neutrino mass" m_β [7],

$$m_\beta = \left[\sum_i |U_{ei}|^2 m_i^2\right]^{\frac{1}{2}} = \left[c_{13}^2 c_{12}^2 m_1^2 + c_{13}^2 s_{12}^2 m_2^2 + s_{13}^2 m_3^2\right]^{\frac{1}{2}} . \quad (1)$$

The second observable — if neutrinos are Majorana spinors — is the effective "Majorana neutrino mass" $m_{\beta\beta}$ in $0\nu\beta\beta$ decay [9],

$$m_{\beta\beta} = \left|\sum_i U_{ei}^2 m_i\right| = \left|c_{13}^2 c_{12}^2 m_1 + c_{13}^2 s_{12}^2 m_2 e^{i\phi_2} + s_{13}^2 m_3 e^{i\phi_3}\right| , \quad (2)$$

where $\phi_{2,3}$ are additional unknown parameters (Majorana phases) [10]. Note that nuclear uncertainties might complicate the interpretation of possible future $0\nu\beta\beta$ signals [17]. The third observable is the sum of neutrino masses in standard cosmology [8]:

$$\Sigma = m_1 + m_2 + m_3 . \quad (3)$$

The oscillation constraints reported in the previous Section induce strong correlations among the above three main observables

Figure 6 shows such correlations in terms of 2σ constraints (bands) in the planes charted by any couple of the absolute mass observables. Note that the bands in the $(m_\beta,\ \Sigma)$ plane of Fig. 6 are quite narrow, due to the high accuracy reached in the determination of all the oscillation parameters. In principle, precise measurements of $(m_\beta,\ \Sigma)$ in the sub-eV range (where the bands for NH and IH branch out) could determine the mass spectrum hierarchy. In the two lower panels of Fig. 6, there remains a large vertical spread in the allowed slanted bands, as a result of the unknown Majorana phases in $m_{\beta\beta}$, which may interfere either constructively (upper part of each band) or destructively (lower part of each band). In principle, precise data in either the $(m_{\beta\beta},\ m_\beta)$ plane or the $(m_{\beta\beta},\ \Sigma)$ plane might thus provide constraints on the Majorana phases.

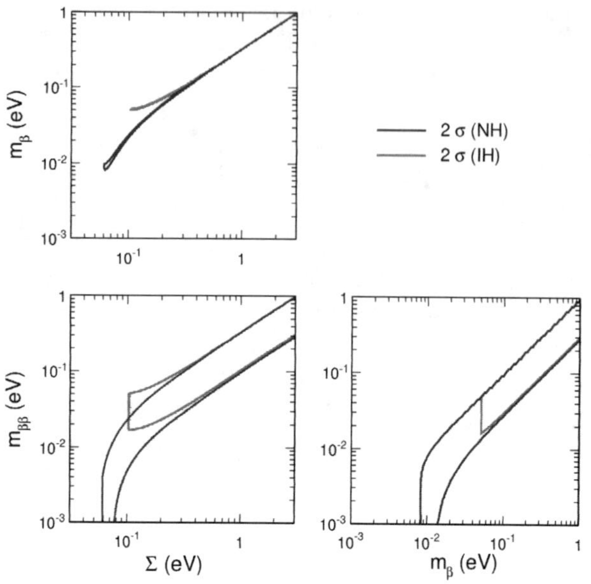

Figure 6. Constraints induced by oscillation data (at 2σ level) in the planes charted by any two among the absolute mass observables m_β (effective electron neutrino mass), $m_{\beta\beta}$ (effective Majorana mass), and Σ (sum of neutrino masses). Blue (red) bands refer to normal (inverted) hierarchy.

At present, there are only safe upper bounds on these absolute mass parameters, at the eV level for m_β [7], and in the sub-eV range for $m_{\beta\beta}$ [9] and Σ [8]. A great experimental activity is in progress towards mass sensitivity goals of $O(\sqrt{\Delta m^2})$, at least via $0\nu\beta\beta$ and cosmological probes. Sensitivities of $O(\sqrt{\delta m^2})$ in $0\nu\beta\beta$ decay appear to be extremely challenging at present.

In the most optimistic scenario, the absolute neutrino masses might be all around 0.1–0.2 eV, and thus observable in the next few years through measurements of at least two among the three (m_β, $m_{\beta\beta}$, Σ) parameters. Then, the concordance of two or three of these observables with the oscillation bands in Fig. 6 woul provide a fundamental cross-check of the standard framework with three massive and mixed neutrinos. If concordance is not achieved (e.g., if strong cosmological limits on Σ are not compatible with possible signals of $m_{\beta\beta} > 0$ within the bands of Fig. 6, or viceversa), the situation would become even more interesting from a phenomenological viewpoint. In this case, data might suggest modifications of the standard framework either in cosmology (e.g., adopting suitable variants of the concordance cosmological model) or in neutrino physics (e.g., exploring nonstandard mechanisms for $0\nu\beta\beta$ decay—a topic witnessing renewed interest). Conversely, the lack of a signal in any of the observables (m_β, $m_{\beta\beta}$, Σ) in the next few years would make the perspectives for the neutrino mass quest extremely challenging.

6. Conclusions

In the light of recent results coming from reactor and accelerator experiments, and of their interplay with solar and atmospheric data (as of December 2013), we have updated the estimated $N\sigma$ ranges of the known 3ν parameters (Δm^2, δm^2, θ_{12}, θ_{13}, θ_{23}), and we have revisited the status of the current unknowns [sign(Δm^2), sign($\theta_{23} - \pi/4$), δ]. The results of the global analysis of all data are shown in Fig. 3 in terms of single parameters. One can appreciate the high accuracy reached in the determination of the known oscillation parameters.

We have also discussed in some detail the status of the unknown parameters. Concerning the hierarchy [sign(Δm^2)], we find no significant difference between normal and inverted mass ordering. However, assuming normal hierarchy, we find possible hints about the other two unknowns, namely: a slight preference for the first θ_{23} octant, and possible indications for nonzero CP violation (with $\sin\delta < 0$), although at a level below $\sim 2\sigma$ in both cases. The second hint appears also in inverted hierarchy, but with even lower statistical significance.

In order to understand how the various constraints and hints emerge from the analysis, and to appreciate their (in)stability, we have considered increasingly rich data sets, starting from the combination of LBL accelerator plus solar plus KamLAND data, then adding SBL reactor data, and finally including atmospheric data. We have discussed the fit results both on single parameters and on selected couples of correlated parameters. It turns out that the hints about the θ_{23} octant appear somewhat unstable at present, while those about δ (despite being statistically weaker) seem to arise from an intriguing convergence of several pieces of data.

Finally, we have discussed the implication of such results for the three observables sensitive to absolute neutrino masses via single- and double-beta decay and cosmology. In general, global analyses of oscillation and non oscillation data appear to provide valuable tools to gauge the overall consistency of the data in a given framework (assumed to be standard 3ν mixing herein). Further experimental data might either confirm the 3ν framework and fix its remaining unknowns (possible CP violation, θ_{23} octant, absolute masses and their ordering, Dirac versus Majorana nature, and Majorana phases in the latter case), or find interesting discrepancies which would require new physics beyond the three known neutrino states and their standard interactions.

Acknowledgments
E.L. is grateful to the organizers, F. Di Lodovico and S. Pascoli, for their kind invitation and hospitality in London, and for their fine organization of a most informative and enjoyable conference. This work is supported in part through the Theoretical Astroparticle Physics Research Programs by the Italian Istituto Nucleare di Fisica Nucleare (INFN TAsP Initiative) and Ministero dell'Istruzione, Università e Ricerca (MIUR PRIN Project).

References

[1] M. Pallavicini, talk at this Conference.
[2] Y. Wang, talk at this Conference.
[3] S. Soldner-Rembold, talk at this Conference.
[4] C. Backhouse, talk at this Conference.
[5] T. Nakaya, talk at this Conference.
[6] F. Capozzi, G. L. Fogli, E. Lisi, A. Marrone, D. Montanino and A. Palazzo, "Status of three-neutrino oscillation parameters, circa 2013," preprint arXiv:1312.2878v1 [hep-ph]. The results shown herein refer to v1 of this preprint, including all 2013 neutrino data. A second version of this preprint [arXiv:1312.2878v2], including T2K data appeared in 2014, has been published on Phys. Rev. D **89**, 093018 (2014).
[7] G. Drexlin, talk at this Conference.
[8] L. Verde, talk at this Conference.
[9] M. Chen, talk at this Conference.
[10] S. Petcov, talk at this Conference.
[11] F. Feruglio, talk at this Conference.
[12] F. Deppisch, talk at this Conference.
[13] J. Kopp, talk at this Conference.
[14] A. Tonazzo, talk at this Conference.
[15] R. Wilson, talk at this Conference.
[16] D. Wark, talk at this Conference.
[17] F. Simkovic, talk at this Conference.

Status of the Prediction of Reactor Anti-neutrino Spectra

Muriel Fallot
SUBATECH (CNRS/in2p3, Univ. of Nantes, EMN) 4, rue A. Kastler, 44307 Nantes FRANCE
E-mail: fallot@subatech.in2p3.fr

Abstract. New generation neutrino physics experiments at reactors have recently determined the value of the θ_{13} mixing angle. Even though their principle is to use multiple detectors allowing to minimize the influence of reactor and nuclear physics ingredients on their results, these ingredients cannot be totally eliminated. They include reactor simulations, but also new computations of reactor anti-neutrino energy spectra. Recently, after a new computation of the reactor anti-neutrino energy spectra, based on the conversion of integral data of the beta spectra from ^{235}U, and $^{239;241}$Pu, a deficit of reactor anti-neutrinos measured by short baseline experiments was pointed out. This is called the reactor anomaly, a new puzzle in the neutrino physics area. Since then, numerous new experimental neutrino projects have emerged. In parallel, computations of the anti-neutrino spectra independent from the ILL data would be desirable. One possibility is the use of the summation method, summing all the contributions of the fission product beta decay branches that can be found in nuclear databases. Studies have shown that in order to obtain reliable summation anti-neutrino energy spectra, new nuclear physics measurements of selected fission product beta decay properties are required. Lately, the first integral measurement of the beta spectrum associated to fast fission of ^{238}U has been performed. Even more recently, the question of the influence of forbidden decays in the determination of reactor anti-neutrino energy spectrum has been raised. At this conference, we will present the methods used to compute reactor anti-neutrino energy spectra, the recent published developments on the topic, remaining open questions and some experimental outlooks.

1. Introduction

Since the last 15 years research in neutrino physics has been very prolific in results, with the determination of most of the oscillation parameters, with the exception of the θ_{13} mixing angle and the δ_{CP} phase. Up until 2011 only an upper bound for θ_{13} was given by the Chooz experiment [1]. Modern accelerator and reactor experiment could tackle the measurement of θ_{13} in a complementary way. Experiments using reactor anti-neutrinos can perform a clean measurement independently of most of the other parameters, and independently from the δ_{CP} value, while experiments using accelerator-produced neutrinos and anti-neutrinos can bring information on both parameters. Indeed in 2011, Minos [2] and T2K [3] provided the community with indications of a non-zero θ_{13} mixing angle through appearance of electron neutrino events in their detectors. The first reactor experiment to release results was Double Chooz, in November 2011 [4]. These results were followed by evidences by Daya Bay [5] and [6] in early 2012. The principle of the new generation reactor anti-neutrino experiments quoted above is to use a detection site near reactor cores to measure the emitted anti-neutrino spectrum at a location where the oscillation has not yet occurred, while a far site will measure the spectrum at the oscillation maximum.

The two sites contain one or several identical detectors so as to cancel most of the detection systematic errors. With the advent of new generation reactor anti-neutrino experiments, re-investigation of reactor anti-neutrino spectra was carried out during the last years. Indeed even with multi-site experiments, one cannot make a measurement fully independent from a spectrum prediction unless having detection sites filling the isoflux condition. With nuclear power plants containing six reactors such as Daya Bay and Reno this condition is not possible to fulfil. In the case of the Double Chooz experiment, the first phase of the experiment takes data with its far detector only and thus its results arise from the comparison of the measurement to a prediction of the anti-neutrino spectrum emitted by the Chooz reactor cores. In its second phase, with the two reactors of the plant in operation, a correction due to non-isoflux conditions will be required as well. The recent re-investigation of the reactor anti-neutrino spectra shed the light on the "reactor anomaly" [7], a deficit in the reactor anti-neutrino flux evidenced in a reanalysis of the short baseline ($\leq$100 m) experiments at reactors, due to an upward shift of the normalization of the newly converted anti-neutrino spectra [8, 9]. Several explanations could be invoked, and neutrino physicists have started to chase the existence of sterile neutrinos in the $\Delta m^2 \simeq 1eV^2$ region with new experimental projects ; at research reactors, but also accelerators and with intense radioactive sources [10]. Nuclear physicists have started to investigate possible uncertainties in the calculation of reactor anti-neutrino spectra, and alternative calculations using nuclear data could be desirable, being independent from the unique measurements of integral beta spectra at ILL in the 80s [11] on which the newly converted spectra rely. In this paper, we will review briefly the actual two main methods employed to compute reactor anti-neutrino spectra, and summarize the on-going discussions regarding these calculations.

2. New Converted Spectra

Taking advantage of the numerous information present nowadays in the nuclear databases [8], the first re-computation of the reactor antineutrino energy spectra relied on the use of nuclear data, with the so-called summation technique (Summation Method, called here "SM"), which will be detailed in the next section. The obtained beta energy spectra for the main fissible isotopes were compared with the integral beta spectra precisely measured by Schreckenbach's team at the ILL research reactor [11]. The comparison showed an agreement at the 10% level, which was not sufficient for neutrino physics experiments. The high energy part of the spectra is over-estimated by the SM and some shape distortions are observed. This trend is compatible with the so-called Pandemonium effect [12]. Interested readers can refer to [8, 13] for further explanations. Benefiting from the gathered nuclear data, the conversion method ("CM") of the ILL beta integral data was then revisited in order to obtain more precise antineutrino spectra. The conversion procedure consisted originally in fitting the integral beta energy spectra with 30 effective beta branches in order to convert them. The new method used most of the experimental beta branches given by the nuclear databases and fitted 90% of the ILL beta spectra with them, fitting the 10% remaining part with 5 effective branches [8]. Thanks to this new method, the correction to the Fermi theory could be applied at the branch level and the real charge (Z) distribution of the fission products could be taken into account, replacing the mean Z fit that was not accurate enough. These corrections led to a 3% normalisation shift upward w.r.t. the antineutrino spectra obtained in [11], triggering the "reactor anomaly" quoted above [7]. Huber revisited the CM in parallel and found similar results [9]. Several possible explanations can be invoked to explain the newly observed deficit of reactor antineutrinos at experiments placed closer than 100m away from reactor cores. There is a large uncertainty associated to the weak magnetism term entering in the computation of beta decay spectra, which could change the normalization of the predictions of the detected antineutrino flux if revealed too far from the adopted value [9]. In addition, these converted spectra rely entirely on the normalization of the ILL data, for which no other measurement exist. These data are still nowadays the data of reference for reactor neutrino experiments. The third potential explanation is the presence of short baseline oscillations, with the mixing of the three standard neutrino flavours into one or two sterile neutrinos [7, 10]. In this context, new measurements or alternative methods to compute the antineutrino energy spectra would be desirable.

3. On the Nuclear Data Side: Summation Method

The anti-ν spectrum associated with one of the 4 fissioning isotopes in a moderated reactor can be computed as the sum of the contributions of all fission products thanks to the use of the full information available per nucleus in Nuclear Databases (NDB). The $\beta/\bar{\nu}$ spectrum per fission of a fissible isotope $S_k(E)$ can be broken-up into the sum of all fission product $\beta/\bar{\nu}$ spectra weighted by their activity A_{fp}

$$S_k(E) = \sum_{fp=1}^{N_{fp}} A_{fp} \times S^b_{fp}(Z_{fp}, A_{fp}, E^b_{0fp}, E) \tag{1}$$

Eventually, the $\beta/\bar{\nu}$ spectrum of one fission product is the sum over the b branches of all β decay spectra (or associated $\bar{\nu}$ spectra), S^b_{fp} (in eq 1), of the parent nucleus to the daughter nucleus weighted by their respective branching ratios. To calculate these quantities, we used the MCNP Utility for Reactor Evolution (MURE) code [14, 21] to extract the fission rates and the fission product activities as a function of time. The Fermi theory has been implemented in MURE in order to calculate the $\beta/\bar{\nu}$ spectrum, S^b_{fp}, of each b branch of the fpth fission product. BR^b_{fp}, Z_{fp}, A_{fp} and E^b_{0fp} have been extracted from the selected NDB.

The β spectrum S^b_{fp} in eq. 1 is derived from the Fermi theory taking into account Huber's prescriptions for the treatment of corrections to be applied to the calculation [9]. In these corrections, two of them should be underlined here for the sake of the understanding of present uncertainties associated to anti-neutrino spectra. We have mentioned in the previous paragraph the weak magnetism correction. This correction arises from the fact that the decaying quark is not free but bound in the nucleon. The weak magnetism correction acts on the spectral shape as a slope as a function of energy, and its average value can be deduced from the magnetic dipole M1 gamma decay width in isobaric analogue states, under the CVC hypothesis. Unfortunately, the required nuclear structure knowledge is limited to a few nuclei, the majority of which being lighter than fission fragments. The representativity of this average value can then be questioned, that is the reason why the weak magnetism correction term is a major source of theory uncertainty. Another source of uncertainty affecting potentially the shape of anti-neutrino spectra is the type of the beta transition. Beta decay follow angular momentum selection rules that can be found in numerous textbooks. We just recall here briefly the main definitions for clarity sake. The beta decay transitions are classified in allowed and forbidden transitions for respectively a zero and a non-zero angular momentum taken by the electron-anti-neutrino pair. If the spin of the electron-anti-neutrino pair is one, the transitions are called Gamow-Teller transitions, while if it is zero they are called Fermi transitions. The forbidden transitions have a weaker rate than allowed ones. In forbidden transitions, the degree of forbiddeness is given by the value of the angular momentum, L=1 means then a first forbidden transition. First forbidden transitions can be a mix of Fermi and Gamow-Teller transitions, except the one corresponding to a total angular momentum, sum of the angular momentum and of the spin of the pair, equal to two. In this case, these transitions can be only of Gamow-Teller type, they are thus called first forbidden unique transitions. These definitions apply to higher values of the angular momentum of the pair. The shape of forbidden transitions beta energy spectra are affected by energy dependent transition nuclear matrices. The forbidden unique transition beta spectral shapes are well determined, while the non-unique ones require to compute the nuclear matrix elements with microscopic nuclear structure models on a case by case basis. We will see in the next paragraph, that forbidden non-unique transitions entering in the composition of the reactor anti-neutrino spectrum, could be an additional source of uncertainty [15].

The last ingredient required to compute our calculation with the SM are nuclear data bases which contain all the nuclei information. We have determined a selected choice of NDB. Details can be found in [16, 13]. The choice was carefully made in order to avoid as much as possible Pandemonium-biased data. The Total Absorption Spectroscopy technique is an alternative to the use of Ge detectors allowing to circumvent the Pandemonium effect [17, 18].

The final calculated spectra obtained with the SM for $^{235;238}$U and $^{239;241}$Pu with 100 keV bins and with the last TAS data can be found in [13]. They are the last up-to-date predictions for these spectra which offer a more precise shape than before [11]. A comparison with the last Huber's computations over the range 2 to 8 MeV has also been proposed. It shows reasonable agreement in the normalization and shape. We reach at maximum 10 % discrepancy up to 7 MeV.

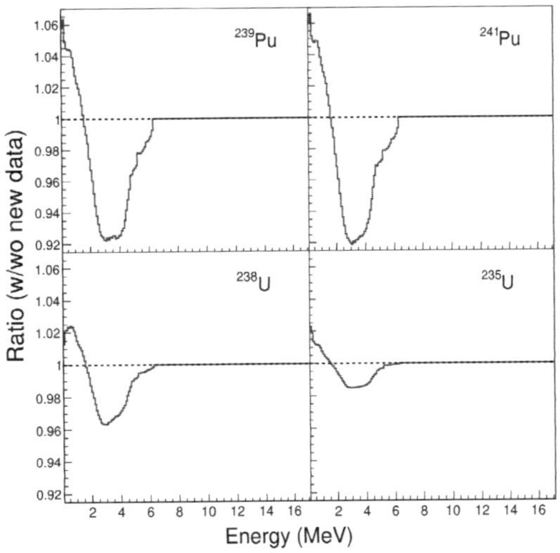

Figure 1. Ratios of the new predicted spectra obtained with and without the last TAS data presented in [17] and extracted from [13].

In Fig. 1, the ratio of the new predicted spectra and the ones obtained with the same data set but the latest TAS data for $^{102;104;105;106;107}$Tc, ^{101}Nb and ^{105}Mo published in [17], previously simulated with JEFF are presented and illustrate how strongly the Pandemonium effect affects the global anti-ν energy spectra of $^{235;238}$U and $^{239;241}$Pu. A noticeable deviation from unity (maximum 8% decrease) is obtained for $^{239;241}$Pu. We observe a maximum of 3.5% deviation for ^{238}U. For ^{235}U, as these nuclei have a small contribution to its spectrum, the effect is smaller than 1.5% at $\sim 3MeV$. Moreover, by convolution with the detection cross section (inverse beta decay), the resulting $\bar{\nu}$ flux corresponding to pure fissions of ^{235}U, ^{239}Pu, ^{241}Pu and ^{238}U is 99,1%, 94.53%, 94.76% and 98.09% respectively relative to previously obtained flux. It stresses again the importance of the Pandemonium effect and the necessity to multiply TAS data measurements to improve the predictiveness of the method. This is emphasized by the fact that the summation method is the only predictive method that can be used to compute anti-neutrino energy spectra associated to any reactor fuel type. These predictions are necessary if one wants to study the potential use of anti-neutrino detectors for reactor core monitoring.

Indeed neutrinos exhibit unique features that have been underlined by the IAEA [?] and that could make of their detection a potential new safeguards tool. Several groups worldwide carry out detector developments to this purpose [20], as well as dedicated reactor simulations [21].

4. Recent Developments

Recently the integral beta energy spectrum arising from ^{238}U fast fission was determined by Haag *et al.* [22] relatively to the ILL ^{235}U energy spectrum from [11]. In this way, the systematic errors of the new measurements are strongly reduced, while the obtained ^{238}U spectral shape is strongly correlated to the ^{235}U beta energy spectrum measured with the BILL spectrometer. After a simple conversion procedure, the anti-neutrino spectrum is given in 250 keV bins in the range from 2.875 MeV to 7.625 MeV with an energy-dependent error of 3.5 % at 3 MeV, 7.6 % at 6 MeV and $\geq$14 % at energies $\geq$7 MeV (68 % confidence level). Furthermore, an energy-independent uncertainty of 3.3 % due to the absolute normalization is added. Compared to the generally used summation calculations, the obtained spectrum reveals a slight spectral distortion of 10 % but returns the same value for the mean cross section per fission for the inverse beta decay. One should note that this new measurement is a major outcome for the determination of reactor anti-neutrino energy spectra, as up until now, the only determination of the ^{238}U anti-neutrino energy spectrum was relying on calculations. On the neutrino physics side, the result is not significantly different enough from the calculations previously used to question the present neutrino physics results like the reactor anomaly or the Double Chooz phase one results on θ_{13}. One should notice that as the new ^{238}U spectral measurement has been performed relatively to the ILL ^{235}U one, its absolute normalization is fully correlated with the one of the latter spectrum, and thus cannot confirm or infirm the normalization of ILL spectra, which is a strong ingredient of the reactor anomaly, uncorroborated by any measurements as they are unique up until now. In addition, the remarkably good agreement [23] found between the measurement and the summation method spectra from [8, 13] gives confidence to a certain extent on the nuclear physics ingredients entering in summation method spectra.

The agreement is not good enough, though, to eliminate all sources of nuclear physics uncertainties affecting the shape of anti-neutrino spectra. Indeed, in addition to the Pandemonium bias affecting some of the important nuclear data entering in the calculation of anti-neutrino spectra, some other nuclear physics uncertainties could affect both the spectra built with the conversion and the summation methods. Recently, a publication by Hayes et al. [15] pointed out that out of $\simeq$ 6000 beta decay transitions, about 1500 are forbidden transitions. Phase space factor and Weak Magnetism expressions have to be modified for Forbidden non-unique transitions, and the corrections are not the same for beta and antineutrinos. Even the converted spectra in which the beta spectrum shape is fixed to the ILL data, could be affected if the Forbidden non-unique transitions happened to be contributing importantly to the anti-neutrino spectra depending on which operator among the possible ones dominate the transitions. Without detailed nuclear structure information there is no method of determining which operators determine the 1500 forbidden transitions. If significant contribution for forbidden non-unique transitions is found in the decays of the fission products, Hayes et al. say that the uncertainty associated to the conversion may be greater than 5% over the whole energy range. Ratios of antineutrino spectrum to the original ILL spectrum allowing different operators to dominate the non-unique forbidden transitions were computed showing that the forbidden transitions introduce an operatordependent distortion of spectrum. In reality there will be a mixture of these and other operators.

Recently, a workshop was organized in Seattle in Nov. 2013 [24] in order to gather the nuclear physics community to discuss about the questions raised by these recent nuclear data studies, in the context of the reactor anomaly. In the following, we give a few outcomes of the workshop. As regards converted spectra, the largest uncertainty [9] comes from the weak magnetism term as explained in the previous sections. Here we could expect that some constraints will come from precised reactor antineutrino experiments (DB, DC, Reno) with a precise measurement of the reactor anti-neutrino energy spectrum shape with their near detectors. This will depend on the systematic error achievable by these experiments on their detection energy scale, since an error in the energy scale acts as a slope on the energy spectrum, exactly like the weak magnetism term. It was noted during the workshop that large log(ft) contribute importantly to the spectra (30%) but we don't know yet how many of them are forbidden non-unique transitions, nor the spin/parity of the transitions. A survey is thus required from nuclear physicists. Potential effects of compensation that would mitigate the impact of these transitions in the full conversion of the ILL spectra were also not computed up until now. An agreement on the expressions of the corrections to Fermi theory should also be found in the following months, as several expressions have been found in the literature [25]. Agreement on a short list of important contributors [27] was found to re-measure to correct from the Pandemonium bias in the databases is established, experiments are on-going in Europe and in the US [26, 27]. The Pandemonium bias distorts largely the Summation Method spectra which need to be corrected before using them, or should be used with a estimate of their present error bars. A survey of forbidden non-unique cases for which the shape of the beta spectra were measured will be performed by nuclear physicists. The status of fission yields needs to be assessed too [28]. Overall the good agreement with Nils Haags measurement of ^{238}U integral beta energy spectrum [22] gives confidence on the present calculations performed with nuclear data.

5. Conclusions

The improvement of the calculations of reactor antineutrino energy spectra is still a challenge for both fundamental and applied neutrino physics. The ILL data are up until now the only and most precise measurements, considered as a reference in neutrino physics. The newly converted anti-neutrino spectra exhibit a normalization shift w.r.t previous anti-neutrino spectra which led to the reactor anomaly [7]. Independent evaluations of the reactor spectra could provide new constraints on the existence of light sterile neutrinos. A possible alternative could be to compute these spectra with the summation method, based on the nuclear data of the fission products nuclear properties. It was shown that the Pandemonium nuclei play a major role in the estimate of the anti-neutrino spectra using the summation method and that TAS measurements of these nuclei could allow us improving drastically the predictiveness of these spectra. An experimental program is on-going with this objective. The possible strong influence of first forbidden non-unique beta transitions on the shape of converted anti-neutrino spectra was also evoked. The inclusion of the shape distortion due to first forbidden non-unique transitions should be tested in the ILL data conversion procedure before drawing any definite conclusions. An effort to provide a summary of the situation regarding these spectra has been launched [24], gathering the neutrino and nuclear communities. The determination of reactor anti-neutrino spectra has a wide interest as it is also mandatory for using anti-neutrino detection for reactor monitoring [19] for safeguards purposes.

6. References

[1] M. Apollonio et al., Chooz collaboration, Eur. Phys. J, C 27, 331.
[2] P. Adamson *et al.*, Phys. Rev. Lett. **107**, 181802 (2011).
[3] K. Abe *et al.*, Phys. Rev. Lett. **107**, 041801 (2011).
[4] Y. Abe *et al.* (Double Chooz collaboration), Phys. Rev. Lett. 108, 131801, (2012).
[5] F. P. An *et al.*, Phys. Rev. Lett. 108 (2012) 171803.
[6] J. K. Ahn *et al.*, Phys. Rev. Lett. 108 (2012) 191802.
[7] G. Mention *et al.*, Phys. Rev. **D83**, 073006 (2011).
[8] Th. Mueller *et al.*, Phys. Rev. **C 83**, 054615 (2011).
[9] P. Huber, Phys. Rev. **C 84**, 024617 (2011). Errata: Phys. Rev. **C 85**, 029901(E) (2012).

[10] K. N. Abazajian et al., http://arxiv.org/abs/1204.5379.
[11] A. A. Hahn *et al.*, PHYS. LETT. **B218**, 365 (1989) and references therein.
[12] J. C. Hardy *et al.*, Phys. Lett. **B71**, (1977) 307.
[13] M. Fallot *et al.*, Phys. Rev. Lett. **109**, 202504 (2012).
[14] MURE, `http://www.nea.fr/tools/abstract/detail/nea-1845`. O. Meplan, Tech. Rep. LPSC 0912 and IPNO-09-01 (2009).
[15] A. Hayes et al. *Preprint* arXiv:1309.4146v2.
[16] M. Estienne *et al.*, International Conference on Nuclear Data for Science and Technology (ND2013), New York, US, to be published in Nuclear Data Sheets Journal (2013).
[17] A. Algora *et al.*, PHYS. REV. LETT. **105**, 202501 (2010).
[18] A.-A. Zakari-Issoufou *et al.*, proceedings of the IEEE int. conf. (2013), Seoul, Corea, to be published. A.-A. Zakari-Issoufou *et al.* EPJ Web of Conferences **62**, 07007, (2013) Fission 2013 - Fifth International Workshop on Nuclear Fission and Fission Product Spectroscopy.
[19] IAEA Report SG-EQGNRL-RP-0002 (2012) and IAEA Report STR-361, (2009).
[20] M. Fallot, International Conference on Nuclear Data for Science and Technology (ND2013), New York, US, to be published in Nuclear Data Sheets Journal (2013).
[21] S. Cormon *et al.*, International Conference on Nuclear Data for Science and Technology (ND2013), New York, US, to be published in Nuclear Data Sheets Journal (2013).
[22] N. Haag *et al.*, Phys. Rev. Lett., **112**, 12 (2014).
[23] INT workshop on reactor neutrinos, Seattle, Nov. 2013: `http://www.int.washington.edu/PROGRAMS/13-3/workshop.html`. See presentation by N. Haag.
[24] INT workshop on reactor neutrinos, Seattle, Nov. 2013 `http://www.int.washington.edu/PROGRAMS/13-3/workshop.html`.
[25] INT workshop on reactor neutrinos, Seattle, Nov. 2013: `http://www.int.washington.edu/PROGRAMS/13-3/workshop.html`. See presentations by P. Vogel, P. Huber and A. Hayes.
[26] Proposal to the PAC of the JYFL, Jyväskylä : Study of nuclei relevant for precise predictions of reactor neutrino spectra. Spokespersons: M. Fallot, A. Algora, J.-L. Tain.
[27] INT workshop on reactor neutrinos, Seattle, Nov. 2013: `http://www.int.washington.edu/PROGRAMS/13-3/workshop.html`. See presentations by A. Sonzogni, E. Mc Cutchan, T. Johnson and M. Fallot.
[28] INT workshop on reactor neutrinos, Seattle, Nov. 2013: `http://www.int.washington.edu/PROGRAMS/13-3/workshop.html`. See presentation by R. Mills.

Results from MINOS and NOνA

C Backhouse

California Institute of Technology
M/C 356-48, 1200 E California Blvd, Pasadena, CA 91125, USA

E-mail: bckhouse@caltech.edu

Abstract.
The MINOS experiment, operating in the NuMI beam since 2005, has provided the most precise measurement of the atmospheric mass splitting $|\Delta m^2_{32}|$, and the recent combination of the ν_μ, ν_e, and atmospheric neutrino samples has provided some evidence of non-maximal mixing, and hints about the neutrino mass hierarchy and the θ_{23} octant. Construction of the NOνA experiment, situated off-axis in the upgraded NuMI beam, is almost complete. Over the coming years it will have significant power to probe the questions of the mass hierarchy, θ_{23} octant, and the possibility of $\mathcal{CP}$ violation in the lepton sector. This paper gives an overview of the results from MINOS, and of the sensitivity of the NOνA experiment.

1. Introduction

The MINOS and NOνA experiments are both situated on the NuMI neutrino beam-line. 120GeV protons from the Fermilab Main Injector strike a graphite target, the resulting pions are focussed by two magnetic horns, and decay to produce a beam of primarily muon neutrinos. Observing the energy-dependent disappearance of these muon neutrinos, and subdominant appearance of electron neutrinos provides information about the neutrino oscillation parameters. Both experiments make use of a two-detector design, in which observations by a Near Detector allow measurements of the beam flux, cross-sections and detector effects, while a larger, but otherwise similarly constructed, Far Detector observes the change in the beam composition over a much longer baseline.

2. MINOS

The MINOS detectors are magnetized steel-scintillator tracking sampling calorimeters. They consist of alternating planes of 2.54cm thick steel and 4.1cm$\times$1cm solid scintillator strips. Each scintillator strip contains an embedded wavelength-shifting fibre, read out at both ends by Hamamatsu multi-pixel PMTs. Adjacent scintillator planes are oriented at 90° to each other, allowing for 3D event reconstruction. The detectors are magnetized by a coil to a field of around 1T, allowing for charge identification of muons from curvature.

The MINOS Far Detector is located underground in the Soudan mine, Minnesota, 735km from the NuMI target. It has a total mass of 5.4kton, and has been operating since 2003. The Near Detector is located underground at Fermilab, about 1km downstream from the target. It has a mass of about 1kton, and was completed in 2005. Since 2005, the NuMI beam-line has delivered 10.7×10^{20} protons-on-target (POT) to MINOS in the "low energy" neutrino configuration, and 3.36×10^{20} POT in antineutrino configuration, plus smaller samples in special configurations. In

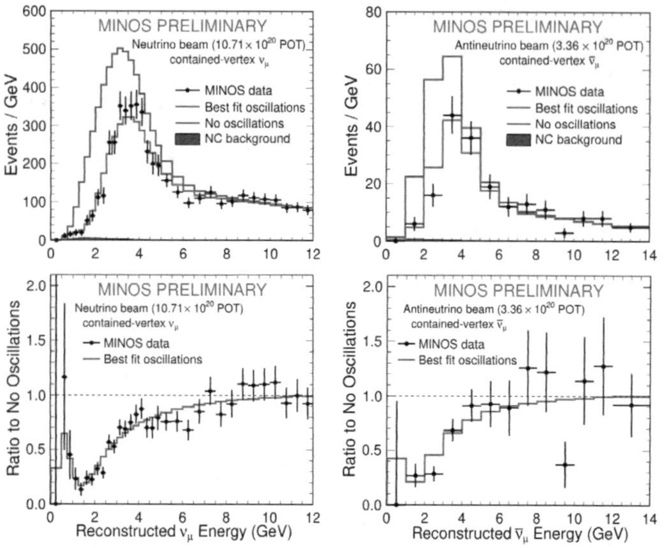

Figure 1. Energy spectra from the MINOS neutrino disappearance analyses. The black points are the observed event counts, with statistical errors. The grey curves show the expectation in the absence of oscillations, and the red curves show the best oscillation fit. Agreement with the oscillation hypothesis is very good for both neutrino mode (left) and antineutrino mode (right). The lower panels display ratios to the unoscillated prediction.

addition, MINOS has collected an exposure of 37.8 kton years of atmospheric neutrinos. In 2013, following a shutdown, the NuMI beam returned, delivering neutrinos to NOνA (and MINOS+ and MINERνA) in the new "medium energy" configuration, optimized to give the greatest flux around oscillation maximum ($\sim$ 2GeV) given NOνA's off-axis location.

Extrapolating from the observed neutrino spectrum in the Near Detector, in the absence of oscillations, MINOS would expect to select 3201 ν_μ interactions in the Far Detector, and 363 interactions in the antineutrino running. The observed counts are 2579 and 312 respectively. Figure 1 shows the spectral shapes of these deficits, compared to the best fit from the neutrino oscillation hypothesis. The three-flavour oscillation paradigm fits the data well: 18% of Monte Carlo pseudo-experiments have a worse fit [1].

The inclusion of atmospheric neutrino samples into the analysis provides additional sensitivity, particularly to the mass hierarchy. The unoscillated expectation is for 1100 events, 905 are observed. Figure 2 shows the spectra of the contained muon-selected events as a function of zenith angle. Partially-contained and showering events are also included in the analysis as additional samples.

Selecting electron neutrino candidates, the expectation for $\theta_{13} = 0$ is 69.1 events of background in neutrino mode, and 10.5 in antineutrinos. The counts observed in data are 88 and 12. Figure 3 shows these excesses in bins of energy and PID value. The hypothesis that $\theta_{13} = 0$ is rejected at the 96% confidence level [2].

Combining the information provided by the ν_μ, atmospheric, and ν_e samples, plus additional constraints on θ_{13} from reactor experiments provides some sensitivity to the neutrino hierarchy and θ_{23} octant. The solar parameters (Δm^2_{21}, θ_{12}) are held fixed in this analysis. θ_{13} is treated as a nuisance parameter, constrained by external reactor results. δ_{CP}, θ_{23}, and Δm^2_{32} are unconstrained and determined by the joint best fit to all the spectra. Major sources of systematic error are included as nuisance parameters.

Figure 4 shows the confidence contours obtained in (Δm^2_{32}, $\sin^2\theta_{23}$) space. The best fit value of $\sin^2\theta_{23}$ is 0.41, and maximal mixing is disfavoured at 79% confidence. The data also weakly favour the inverted hierarchy and the lower θ_{23} octant. Figure 5 shows a different projection of this information, with δ_{CP} information included. The most disfavoured scenario is the combination of normal hierarchy and upper θ_{23} octant which is rejected at the 81% confidence

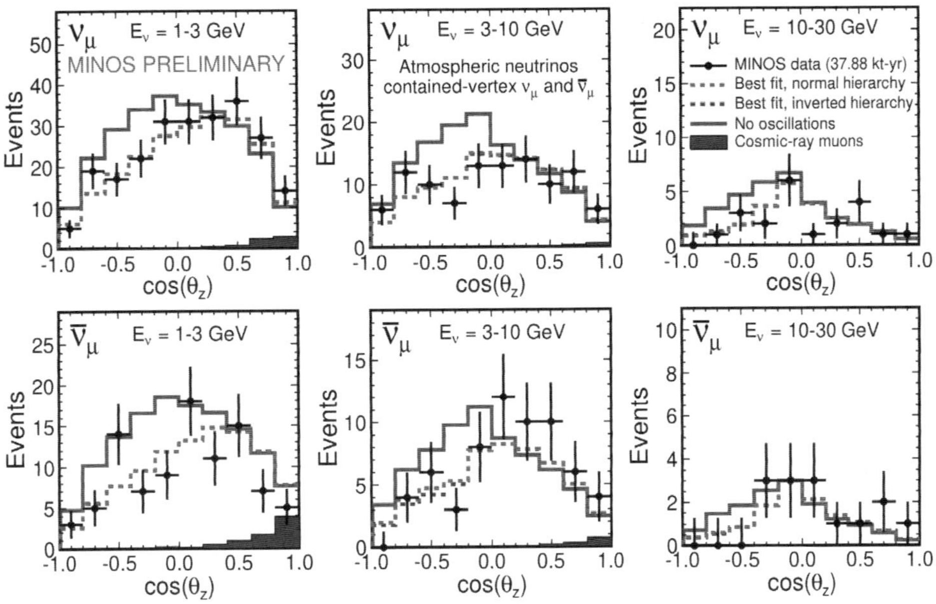

Figure 2. Spectra for the MINOS atmospheric neutrino sample. The points show the observed event counts as a function of reconstructed zenith angle. The grey curves show the unoscillated predictions, and the red and blue dashed curves are the best oscillation fits assuming normal or inverted hierarchy respectively. The filled blue histograms indicate the background due to cosmic-ray muons.

Table 1. Summary of MINOS combined oscillation fit results for normal and inverted hierarchy.

Parameter	Best fit	Confidence limits (NH)	Best fit	Confidence limits (IH)
$\|\Delta m^2_{32}\|/10^{-3}\text{eV}^2$	2.37	2.28 - 2.46 (68% C.L.)	2.41	2.32 - 2.53 (68% C.L.)
$\sin^2\theta_{23}$	0.41	0.35 - 0.65 (90% C.L.)	0.41	0.34 - 0.67 (90% C.L.)

Preference for inverted hierarchy: $-2\Delta\log L$=0.23
Preference for lower octant: $-2\Delta\log L$=0.09
Exclusion of maximal mixing: $-2\Delta\log L$=1.54 (→ 79% C.L.)

level for all values of δ_{CP}. This measurement is also the most precise determination of $|\Delta m^2_{32}|$. The results are summarized numerically in Table 1 [3].

If some component of the disappearing MINOS flux were oscillating to sterile neutrinos, one would expect to also see a deficit of neutral current events, in addition to the deficit seen in ν_μ charged currents. In fact, a search for this effect found a small excess of neutral current interactions at the Far Detector as compared to predictions. Interpreting in terms of oscillations around $\Delta m^2 \sim 0.5\text{eV}^2$ leads to a limit $\sin^2 2\theta_{\mu e} < 7.1 \times 10^{-3}$ at the 90% confidence level.

The MINOS detectors continue to operate in the NOνA-era beam. The higher energy peak of the neutrino flux reduces the statistics acquired in the oscillation region , but the statistics at higher energies are greatly increased. MINOS+ expects to collect around 4,000 ν_μ charged current events per year. This high-statistics sample will allow tests of the standard three-flavour oscillation paradigm, including searches for sterile neutrinos, and sensitivity to signatures of

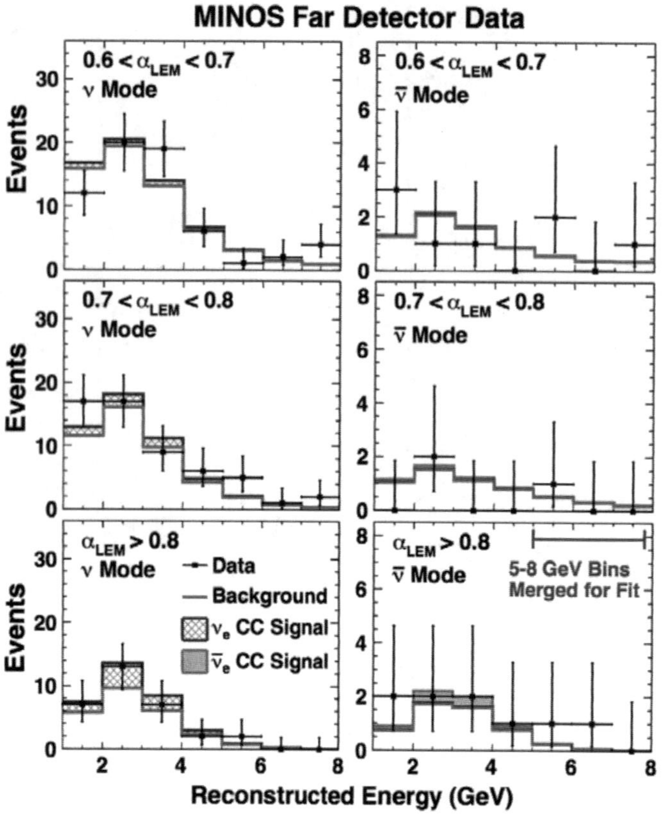

Figure 3. Spectra of ν_e-selected events in MINOS. The points are the number of events selected as a function of reconstructed neutrino energy. The red curve shows the prediction for $\theta_{13} = 0$, and the purple region shows the excess for the best-fit value of θ_{13}.The samples are divided by PID value (top to bottom) and between neutrino (left) and antineutrino (right) running. The data is divided into bins of PID value, from the lowest selected PIDs (top) to the most ν_e-like events (bottom). 88 neutrino-mode events and 12 antineutrino-mode events are selected, compared to an expectation for $\theta_{13} = 0$ of 69.1 and 10.5. The expectations for $\theta_{13} = 0.1$ are 95.1 and 13.6.

exotic physics such as large extra dimensions.

3. NOνA

The NOνA detectors are fine-grained, low-Z, highly active tracking calorimeters. Cells of extruded PVC, approximately 6cm×4cm are formed into planes, with the orientation of successive planes alternating as in MINOS. They are filled with a mineral oil/liquid scintillator mix, which comprises 64% of the detector by mass. Each cell contains a looped wavelength-shifting fibre, read out at one end by one of the 32 pixels of a Hamamatsu avalanche photodiode (APD), cooled to $-15°$C [4].

The Far Detector is located on the surface at Ash River, Minnesota, 810km from the NuMI target. The detector consists of 344,000 channels, approximately $16 \times 16 \times 60$m, for a total mass of 14kton. Assembly and placement of the plastic blocks, scintillator filling, and electronics installation proceed in parallel. All blocks are now in place, and oil filling is almost complete. Over 2/3 of the detector is instrumented with APDs and being read out. The Near Detector is located underground at Fermilab, approximately 1km from the target. It comprises 18,000 channels, with a mass of 0.3kton. All blocks are in place, and oil filling and electronics outfitting are in progress. The NuMI beam is being upgraded from a maximum capacity of 350kW up to 700kW. This will be achieved by converting the Recycler from running antiprotons to protons, and shortening the cycle in the Main Injector from 2.2s to 1.33s. The beam has been running stably since returning from shutdown in September 2013, and power is gradually being increased. The maximum power achievable will be limited to 500kW until upgrading of the Booster RF

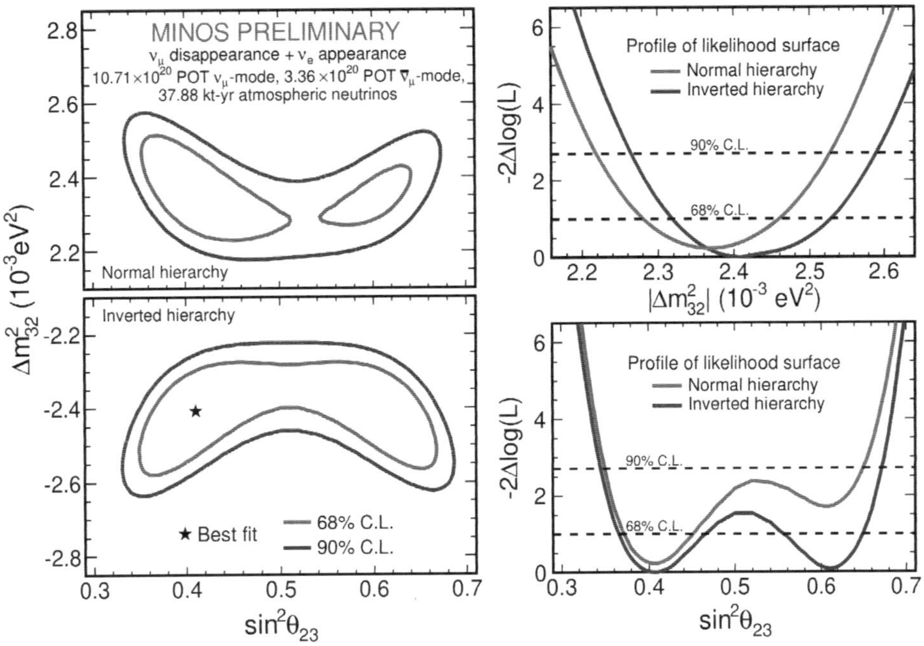

Figure 4. Combined results from the MINOS ν_μ, ν_e, and atmospheric analyses. The star shows the best fit values of $\sin^2\theta_{23}$ and Δm^2_{32}. This point weakly favours non-maximal mixing, the lower octant, and inverted hierarchy. The red and blue contours indicate 68% and 90% confidence levels. The panels on the right display the profile likelihood projection onto each of the two parameters.

cavities is complete. Thus far, in excess of 1.5×10^{20} protons have been delivered.

It is important to validate the detectors' ability to pick neutrino interactions out of the cosmic-ray background, and to verify the synchronization of the Far Detector to the NuMI spill times. Analysis of the Near Detector prototype on the surface at Fermilab (NDOS) shows a clear peak in the neutrino candidate timing spectrum over the cosmic-ray background. Propagating this timing information to the Far Detector gives a 10μs window in which neutrinos are expected to be found. A combination of automated analysis and hand-scanning of events in a 500μs window around this point (plus out-of-time spills for background estimation) found four candidate events within the timing window, on a background of 0.05. The times of the events relative to the beam window were kept secret until after the choice of these events had been finalized. Figure 7 shows the NDOS timing peak compared to the times of the selected neutrino candidates, and Figure 8 shows one of the Far Detector events in the timing window, a multi-pronged interaction, with reconstructed neutrino direction coming from Fermilab.

The sensitivity projections presented below assume 18×10^{20} POT of data taken in neutrino mode plus 18×10^{20} POT taken with antineutrinos. This corresponds to a nominal 6×10^{20} POT per year and a run plan with three years in each mode. The choice of what exposure to take in each beam configuration is flexible, and may be adjusted in response to changing circumstances. The oscillation parameters assumed are $\sin^2 2\theta_{13} = 0.095$ $\sin^2 2\theta_{23} = 0.95$ or 1.0 depending on context. We take $\Delta m^2_{32} = 2.35 \times 10^{-3}$eV^2, and the other parameters from the latest Particle Data Group averages [5].

Current global constraints on the atmospheric mixing angle are around $\sin^2 2\theta_{23} \gtrsim 0.9$,

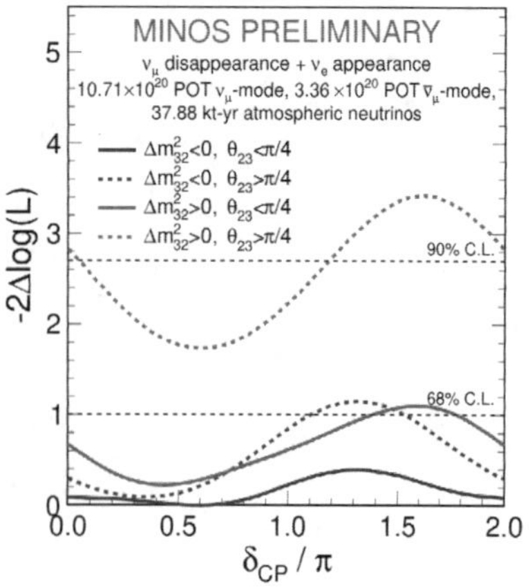

Figure 5. Log-likelihood differences from the best fit oscillation parameters for different values of δ_{CP}, hierarchy and θ_{23} octant for the combination of MINOS analyses. The region normal hierarchy, upper octant, $\delta_{CP} \sim \frac{3\pi}{2}$ is disfavoured at greater than 90% confidence.

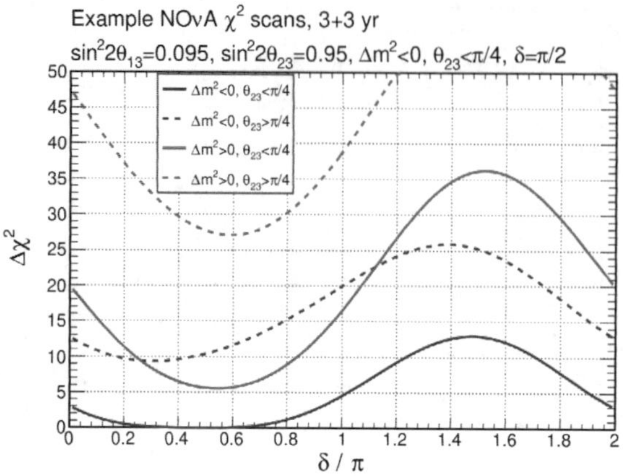

Figure 6. Sensitivity of NOνA for a nominal run of three years neutrino plus three antineutrino (18×10^{20} POT each). The true oscillation values are chosen to be close to the MINOS best-fit. The shape of the curves is very similar to the MINOS results in Figure 5, but the expected sensitivity is much higher. For example, the most disfavoured scenario, normal hierarchy, upper octant, would be excluded at 5σ for all values of δ_{CP}.

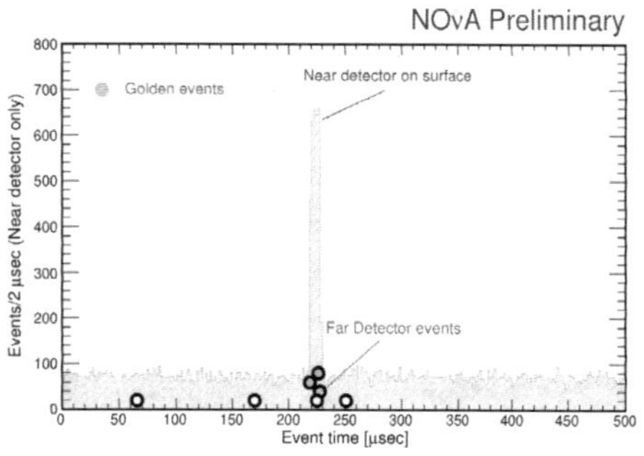

Figure 7. The timing peak formed by neutrinos from the NuMI beam in the NDOS detector (filled histogram) with the times of selected interactions in the NOνA Far Detector marked (circles). The Far Detector events were found by a combination of hand-scanning and automated analysis. The estimated cosmic-ray background in the $10\mu s$ Far Detector spill window is 0.05 events.

with hints of non-maximal mixing from MINOS. The NOνA experiment will have significantly enhanced power to distinguish between maximal and non-maximal mixing. The ν_μ charged current analysis being developed separates the data into three samples: quasi-elastic candidates, which are very pure with good energy reconstruction; the remainder of contained candidates; and uncontained events, where the muon exits through the back or side of the detector, a sample with substantial statistics, but poor energy resolution. Predicted event counts are given in Table 2. The contours that would be obtained with increasing exposure are shown in Figure 9, for two possible values of θ_{23}. With three years of neutrino running, and three years of antineutrinos, this analysis should obtain percent-level uncertainty on the atmospheric mixing parameters. If $\sin^2\theta_{23} = 0.95$, maximal mixing could be excluded at 90% confidence with one year of running.

Table 2 also shows the representative event counts for the sample of electron neutrino candidates. As with MINOS, sensitivity to the mass hierarchy, θ_{23} octant, and δ_{CP} comes from combining the disappearance and appearance analyses, plus constraints on θ_{13} from reactor

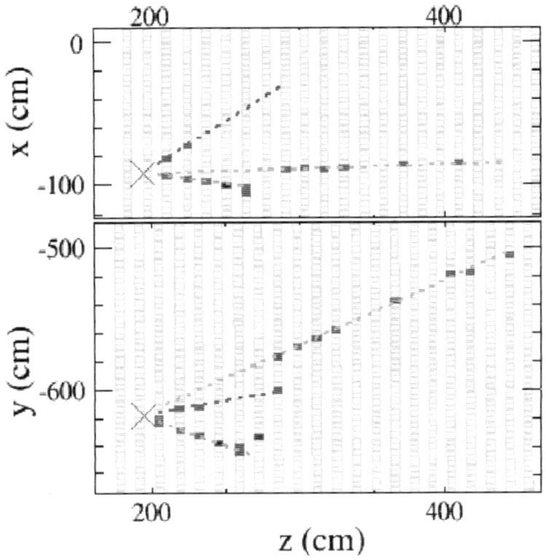

Figure 8. Event display of the first neutrino candidate event selected in the NOνA Far Detector. Each square represents energy deposited and detected in a single cell. The neutrino beam enters from the left. The top panel shows the information from vertically-oriented cells, with horizontal cells shown in the bottom panel. The overlaid coloured lines are the reconstructed paths of particles, and the cross marks the reconstructed position of the interaction vertex.

Table 2. Representative event counts for NOνA analyses. The columns represent 18×10^{20} POT neutrino-mode and antineutrino-mode running. The ν_μ analysis divides the dataset into three samples, candidate quasielastic events, non-quasielastic events, and events where the muon is not fully contained within the detector. The counts are constrained to the range 0-5GeV, the region of sensitivity to oscillations driven by the atmospheric mass splitting.

ν_e **selected**	ν	$\bar{\nu}$
NC	19	10
ν_μ CC	5	<1
Beam ν_e	8	5
Tot bkg	32	15
Signal	68	32

ν_μ **selected**	ν_μ CC	NC	$\bar{\nu}_\mu$ CC	NC
Quasielastic	82	<1	49	<1
non-QE	168	14	78	6
Uncontained	233	6	134	3

(0-5GeV visible energy)

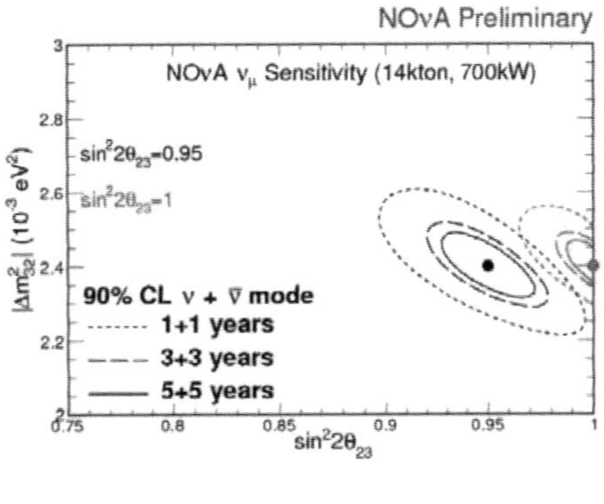

Figure 9. Projected 90% confidence contours in the (Δm^2_{32}, $\sin^2\theta_{23}$) space for two, six, or ten years of running (assuming 6×10^{20} POT per year), for two values of θ_{23}. In the case of $\sin^2 2\theta_{23} = 0.95$, maximal mixing can be excluded at 90% confidence in under two years of running.

experiments. Figure 10 shows the sensitivity of this analysis to these three parameters, as a function of the true values of δ_{CP} and the mass hierarchy. For the mass hierarchy, there are degenerate regions where it is not possible to disentangle the matter effects, which give information about the mass hierarchy, from true $\mathcal{CP}$-violation driven by δ_{CP}. But for favourable combinations $2-3\sigma$ evidence for the mass hierarchy can be obtained. Determination of the θ_{23}

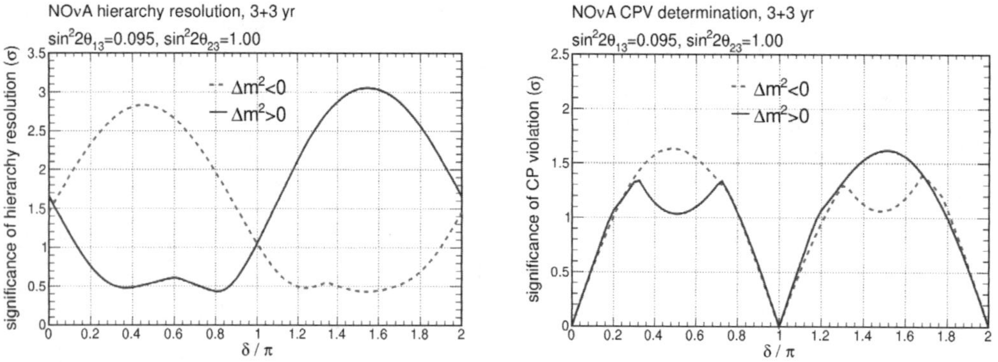

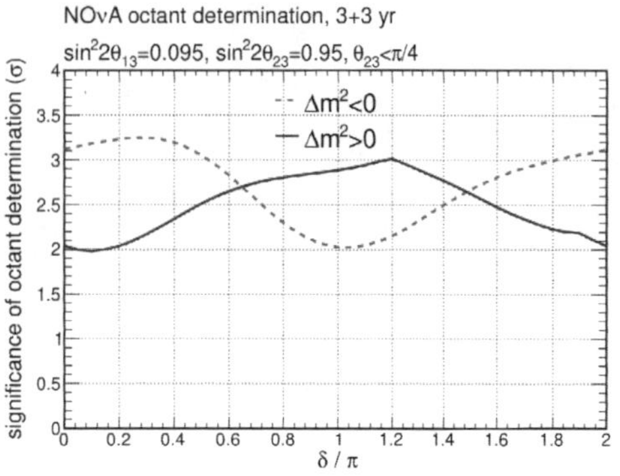

Figure 10. Sensitivity of NOνA to resolve the mass hierarchy (top left), discover $\mathcal{CP}$ violation (above), and resolve the θ_{23} octant (left) as a function of the true value of δ_{CP}. The solid blue curves are in the case that the true mass hierarchy is normal, the dashed red curves are for the inverted case. In each case, $\sin^2 2\theta_{13} = 0.095$ is assumed. For hierarchy and $\mathcal{CP}$ violation $\sin^2 2\theta_{23} = 1$ is assumed, the octant sensitivity assumes $\sin^2 2\theta_{23} = 0.95$, first octant.

octant is much less sensitive to δ_{CP} and can be achieved at better than 2σ for all values of δ_{CP} if $\sin^2 2\theta_{23} = 0.95$. The lower octant is assumed in order to match the best fit from MINOS, performance for the upper octant is slightly better. Figure 6 shows predicted $\Delta\chi^2$ slices as a function of δ_{CP}, to be compared with the analogous Figure 5 from MINOS data. The true parameters are chosen to match the MINOS best fit (inverted hierarchy, lower octant, $\delta_{CP} \sim \frac{\pi}{2}$). If the true parameters are indeed in this region, NOνA would reject parts of phase space (in particular the combination of normal hierarchy and upper octant) at very high significance.

4. Conclusion

The MINOS experiment has collected a large sample of neutrino interactions in over six years of running, in both neutrino and antineutrino mode. A combined fit to this data gives the world's most precise measurement of $|\Delta m^2_{32}|$, and hints for the neutrino mass hierarchy and θ_{23} octant. These detectors will continue taking data in the NOνA-era beam as MINOS+.

Construction of the NOνA experiment is progressing well. First neutrinos have been found in the Far Detector, and projections show good sensitivity to the mass hierarchy and θ_{23} octant, in addition to improved measurements of the atmospheric mixing parameters. Results from the next few years, in combination with other experiments, should provide significant information on these questions for the first time.

References

[1] Adamson P *et al.* 2013 *Phys. Rev. Lett.* **110**, 251801
[2] Adamson P *et al.* 2013 *Phys. Rev. Lett.* **110**, 171801
[3] Adamson P *et al.* 2014 *Preprint* hep-ex/1403.0867, submitted to *Phys. Rev. Lett.*
[4] Ayres D S *et al.* 2007 FERMILAB-DESIGN-2007-01
[5] Beringer J *et al.* 2012 *Phys. Rev. D* **86**, 010001

Statistical issues in long baseline neutrino physics

Alessandra Tonazzo* **for the LBNO Collaboration**

*APC, Univ Paris Diderot, CNRS/IN2P3, CEA/Irfu, Obs de Paris, Sorbonne Paris Cité, France

E-mail: tonazzo@in2p3.fr

Abstract. An animated debate has been ongoing in the neutrino community on how to estimate and quote the expected sensitivity of future long-baseline neutrino experiments to key parameters such as Mass Hierarchy or CP violation. We will present an overview of some items covered by recent papers and will detail the approach chosen by the LBNO Collaboration to present its results.

1. Introduction

Future neutrino experiments using long-baseline beams and very large detectors will address key questions that are still open, such as the ordering of the three known neutrino mass eigenstates (Mass Hierarchy, MH) and the existence of CP violation (CPV) in the leptonic sector. In particular, the LBNO Collaboration [1] has proposed the installation of a liquid argon (LAr) TPC [2] in an extension of the Pyhäsalmi mine in Finland, which will be able to cover important issues in particle physics (proton decay) and astrophysics (neutrinos from supernovae and from the sun). A neutrino beam from CERN, located at a distance of 2300 km, would allow assessment of the MH and exploration of CPV. The experiment envisages a staged approach with two phases: a first phase with 20 kt (fiducial volume) of LAr and a 750 kW beam based on the existing SPS accelerator complex,providing about 1.2×10^{20} p.o.t. per year, and a second phase with a total LAr fiducial mass of 70 kt and a 2 MW beam from a newly-built high-power accelerator (HPPS), providing an intensity 20 times higher.

Since the assessment of CPV depends on the knowledge of the MH, the LBNO strategy is to ensure that MH is known "with high confidence level" in the first few years of running. How to assess this confidence level has recently been at the heart of animated debates.

Papers published in the past couple of years have raised questions about the assumptions on the analytical form of the distribution functions to be used for the computation of the confidence level [3, 4], on how to present the information on sensitivities in a complete and unambiguous way [5] and on the difference between a Bayesian and a frequentist approach [6]. After a short review of some of these topics, we will present the approach chosen by the LBNO collaboration to present its results [7] following the guidelines of [5].

2. Hypothesis test: some basics

The aim of a test of hypothesis is to test a "null hypothesis" H_0 against an "alternative hypothesis" H_1, using a given set of data. The hypotheses are "simple" if they do not contain unspecified parameters, otherwise they are "composite".

For simple hypotheses, the frequentist method proceeds as follows:

Journal of Physics: Conference Series **598** (2015) 012005 doi:10.1088/1742-6596/598/1/012005

1. define a "test statistic" T, which is a function of your data;
2. construct the probability density function (PDF) of T under each hypothesis;
3. define a "critical region" Ω_C such that values of T in Ω_C suggest H_0 to be true: H_0 will be accepted if T falls in Ω_C, otherwise H_1 will be accepted;
4. evaluate the probability to give the wrong answer, which you can do in two ways:
   - reject H_0 when it is true: this is the error "of type I" or "loss", usually denoted as α;
   - accept H_0 when H_1 is true: this is the error "of type II" or "contamination", β.

This is illustrated, for a simple example, in Figure 1.

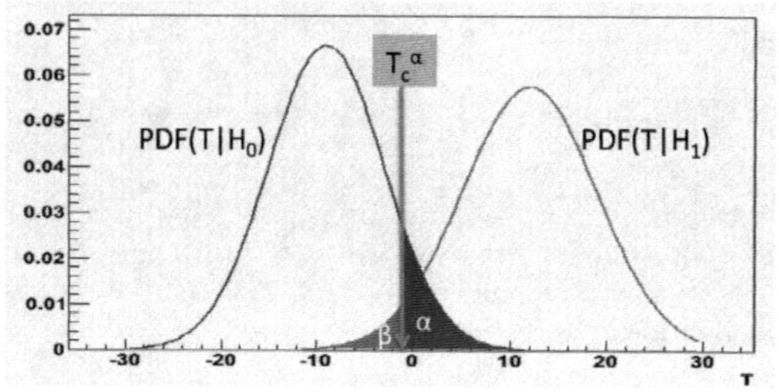

Figure 1. Illustration of the basic ingredients of a hypothesis test in a simple case: PDF of the test statistic T under the two hypotheses, definition of the critical value T_C (the critical region is $T < T_C$), α and β errors.

The "Confidence Level" (CL) for rejecting the null hypothesis is defined as $CL = 1 - \alpha$ and the "Power" to test H_1 against H_0 is $p = 1 - \beta$. For simple 1-dimensional cases such as the one depicted in Figure 1, the critical region is defined by a threshold T_C on the value of T such that H_0 is accepted when $T < T_C$ and rejected when $T > T_C$; the CL and the power for rejection H_0 are given by

$$CL = 1 - \alpha = \int_{-\infty}^{T_C} PDF(T|H_0)\, dT$$

$$p = 1 - \beta = \int_{T_C}^{+\infty} PDF(T|H_1)\, dT$$

It is common practice to use the terminology proper to two-sided Gaussian integrals even if one-sided integrals are actually involved, thus for instance "3σ CL" should be interpreted as $\int_{-\infty}^{T_C} PFD(T|H_0)\, dT = 99.73\%$.

Only if the two values of α and β, or equivalently CL and Power, are quoted is the hypothesis test fully quantified.

Two approaches are frequently used:

(i) set T_C such that $p = 0.5$. The corresponding CL is often referred to as "median sensitivity".

(ii) set T_C such that $\alpha = \beta$, or $CL = p$. This is called "crossing sensitivity" in [5].

They will be discussed below for the case of the MH.

When the hypotheses are composite, the CL and the power depend on some parameters, whose values are in general unknown.

3. Mass Hierarchy as a hypothesis test

The case of mass hierarchy determination can be treated as a test of hypotheses. The two hypotheses are Normal Hierarchy (NH, $\Delta m^2_{32} > 0$) and Inverted Hierarchy (IH, $\Delta m^2_{32} < 0$). Both of them are composite, as in general the distributions of test statistic depend on other oscillation parameters (mixing angles, CP violating phase, etc.) or on detector details. This will be taken into account in the following discussion.

The test statistic T can be constructed thanks to the Neyman-Pearson lemma, which implies that, for simple hypotheses, the most powerful test for any given CL is the likelihood ratio. This can be extended to the case of composite hypotheses. For the MH problem, the likelihood ratio is easily proven to be equivalent to the difference of the χ^2's between data and expectations, where the χ^2 is computed on a single- or multi-dimensional distribution of kinematic variables of the data (such as energy, transverse momentum, etc.), possibly including systematic effects as nuisance parameters. The test statistic T is thus defined as

$$T = \chi^2_{\rm IH} - \chi^2_{\rm NH}.$$

The exact definition of the χ^2 used for the LBNO experiment is provided in [7].

3.1. Test statistic distributions and confidence levels

If H_0 and H_1 are nested hypotheses, i.e., if H_1 depends on n parameters and H_0 can be obtained from H_1 by fixing m of these parameters, Wilks' theorem [8] states that the PDF of T when H_0 is true follows a χ^2 distribution with $n - m$ degrees of freedom. In the past, the CL for the MH problem was often computed assuming that the PDF for a given hierarchy was distributed as a χ^2(1 d.o.f.).

However, as pointed out in [3, 4], the MH problem is clearly NOT a case of nested hypotheses! The correct distributions for the PDF of the test statistic T under the two hypotheses must be evaluated. It can be demonstrated from simple arguments [3, 4] that the two distributions are well approximated by Gaussians:

$$PDF(T|{\rm NH}) = N(|T_0^{\rm NH}|, 2\sqrt{|T_0^{\rm NH}|}) \quad ; \quad PDF(T|{\rm IH}) = N(-|T_0^{\rm IH}|, 2\sqrt{|T_0^{\rm IH}|}).$$

This approximation is verified at the percent level by generating the distribution with a toy MonteCarlo simulation, as shown for the LBNO experiment in Figure 2.

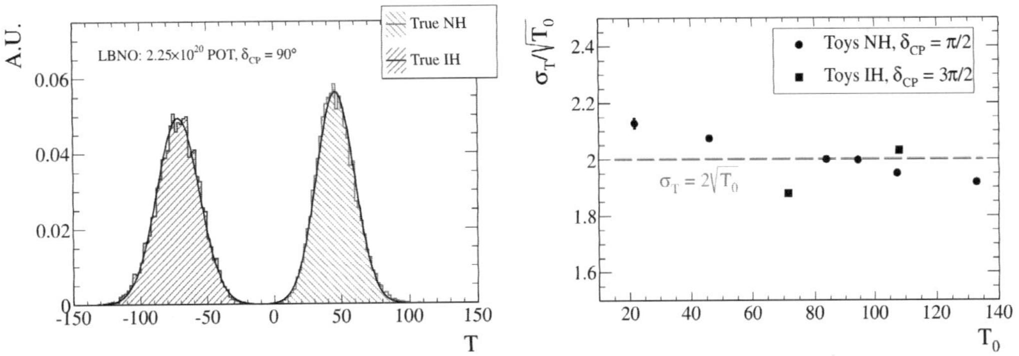

Figure 2. Distribution of the test statistic for the MH test in the LBNO experiment (left) and ratio between width and mean of the distributions as a function of exposure (right).

In the simple approximation $|T_0^{\rm NH}| = |T_0^{\rm IH}|$, which holds for many experiments, it is easy to compute the significance of the test in terms of T_0 with the two prescriptions for CL and power quoted in the previous section. Let us consider, to fix ideas, H_0=IH. Then,

- to have $p = 0.5$, one should set $T_C = T_0^{\rm NH}$ and one gets $CL = N\sigma$ with $N = \sqrt{|T_0^{\rm IH}|}$;
- to have $p = CL$, one should set $T_C = 0$ and one gets $CL = N\sigma$ with $N = \sqrt{|T_0^{\rm IH}|/2}$.

The same conclusion holds with H_0=NH.

What we wish to stress here is the factor 2 that appears between the two cases. This has led to a lot of misunderstanding, which is however fully resolved if the power of the test is quoted along with the CL. We note in passing that the factor 2 would also appear if the "crossing sensitivity" was computed with a PDF having $\sigma = \sqrt{T_0}$, which is however completely wrong for the MH case !

3.2. Construction of a hypothesis test for MH

In order to construct properly a hypothesis test for MH, the LBNO Collaboration has chosen the approach described in [5]. To fix ideas, suppose that the IH is taken as H_0. Define the desired CL for its exclusion and set the critical value $T_C^{\rm IH}$ such that

$$\int_{-\infty}^{T_C^{\rm IH}} PFD(T|{\rm IH})\, dT = CL,$$

then compute the corresponding power as

$$p = \int_{T_C^{\rm IH}}^{+\infty} PFD(T|{\rm NH})\, dT.$$

This can be done for a given value of $T_0^{\rm IH}, T_0^{\rm NH}$ or for all possible values of $T_0^{\rm IH}, T_0^{\rm NH}$: typically $|T_0|$ increases with exposure of the experiment, so the power can also be provided as a function of exposure or time. The procedure should then be repeated with NH as H_0, which in general will yield a different critical value $T_C^{\rm NH}$.

Up to now we have neglected the fact the the hypotheses are composite. The distributions of T depend on additional parameters, related to neutrino mixing or to detector performance. In particular, for the LBNO experiment, they depend on the value of the CP violating phase δ_{CP}, as shown in figure 3. This parameter will not be known at the time of the experiment.

As pointed out in [5], the desired CL must be ensured for all values of the (unknown) parameters, and in general the power will depend on the parameters. For the MH case, one should thus set T_C in the most pessimistic case, i.e. for the smallest absolute value of T_0 in either case. We thus compute

$$T_C^{\rm IH} \text{ for } T_0^{\rm IH}(\delta_{CP} = 3\pi/2)$$
$$T_C^{\rm NH} \text{ for } T_0^{\rm NH}(\delta_{CP} = \pi/2).$$

The power can then be evaluated as a function of exposure for all the possible values of δ_{CP}, as in the plots shown on Figure 4.

From these plots one can easily read the exposure needed to achieve the given CL whatever value of δ_{CP} (lower end of the bands) and with a given power (for example, 0.5 or 1). It is common practice to quote the exposure that provides a 50% chance to achieve the desired CL, however this may be risky when implying the decision to invest time and money in a future experiment. LBNO has chosen to quote its sensitivity with the exposure ensuring a probability close to 100% to achieve the desired CL, i.e. for power close to 1.

Either choice is correct, as long as the power is quoted along with the CL. LBNO has ~100% probability (power ~1) to achieve MH discrimination at 5σ CL, for either ordering, after 4-5 years of its first phase.

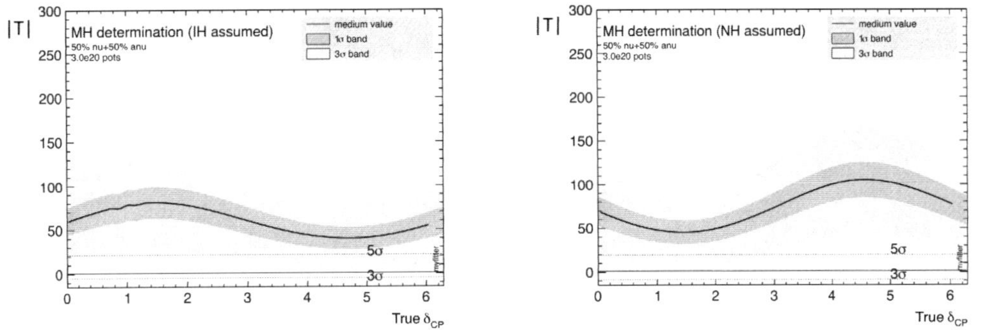

Figure 3. Test statistic values for MH as a function of δ_{CP}, under the hypothesis of IH (left) and of NH (right). The exposure corresponds to about 2 years of running of LBNO 1st phase.

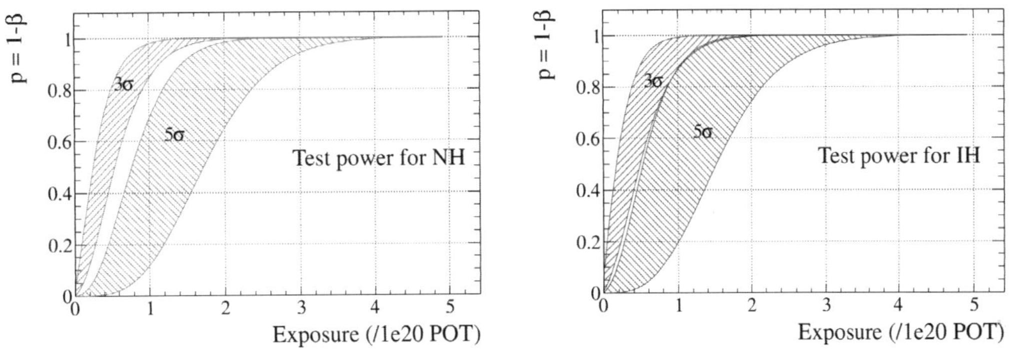

Figure 4. Statistical power as a function of exposure for the test of NH (left) and IH (right) for 3σ and 5σ CL. The shaded bands correspond to the variation of δ_{CP}.

4. The Bayesian approach to MH determination

In the frequentist approach to the MH test described so far, two sets of information must be provided: CL and power for NH and for IH. With a Bayesian approach [6], a single set of values can contain all the information on the test.

In the Bayesian approach, one needs to define a "prior" P on each hypothesis, which effectively provides a relative normalisation of the two PDF's. It can be prove that the most conservative assumption is to consider the two hypotheses to be equally likely, i.e. $P(\mathrm{NH}) = P(\mathrm{IH}) = 0.5$, and we will use this in the following.

One then sets a threshold t and computes two critical values

$$T_C^{\mathrm{NH}} \text{ such that } \frac{PDF(T|\mathrm{NH})}{PDF(T|\mathrm{NH}) + PDF(T|\mathrm{IH})} > t \text{ for } T > T_C^{\mathrm{NH}}$$

$$T_C^{\mathrm{IH}} \text{ such that } \frac{PDF(T|\mathrm{IH})}{PDF(T|\mathrm{NH}) + PDF(T|\mathrm{IH})} > t \text{ for } T < T_C^{\mathrm{IH}}$$

Essentially, t represents the "odds" or ratio of posterior probabilities: if an experiment gives a results T, the probability that nature is actually NH (or IH) is larger than t in the two critical regions. This is a question that it makes no sense to ask in the frequentist approach.

Finally, one can define a "p-value" for the test as

$$p = \int_{-\infty}^{T_C^{\rm IH}} PFD(T|{\rm IH})\; dT + \int_{T_C^{\rm NH}}^{+\infty} PFD(T|{\rm NH})\; dT$$

representing the overall probability to obtain a value of T such that one of the hypotheses has a posterior probability of at least t.

This approach has been taken, for example, in [9] for the study of future large underwater or under-ice detectors.

5. CPV as a test of hypothesis

For the case of CP violation in the leptonic sector, the null hypothesis H_0 is the absence of CPV ($\delta_{CP} = 0$ or π) and the alternative hypothesis is any other value of the CPV phase. The test statistic is defined as

$$\Delta\chi^2 = min(\Delta\chi^2(\delta_{CP} = 0), \Delta\chi^2(\delta_{CP} = \pi))$$

One can refer again to [7] for the exact definition of the χ^2 in the case of the LBNO experiment. The distribution of the test statistic is strongly dependent on the assumptions on oscillation parameters and experimental errors, however the main dependence is by far the one on the value of δ_{CP} itself.

This test of CPV is thus a case of nested hypotheses with one free parameter, δ_{CP}. According to Wilks' theorem [8], the test statistic follows a χ^2 distribution with 1 degree of freedom, and this is confirmed by toy MonteCarlo simulation in the case of the LBNO experiment. An example is shown in Figure 5 (left): this distribution is independent of the exposure, unlike for the MH case.

The PDF of the test statistic under the alternative hypothesis is not easy to predict, it can be estimated with toy MonteCarlos for each value of δ_{CP} and of the exposure. An example is shown in Figure 5 (right). A simple parametrisation with a skewed Gaussian, depending on the exposure, is found to be a good approximation for this distribution. The power of the test can be computed, for each value of δ_{CP}, by integrating numerically or analytically the distribution. The LBNO collaboration has chosen, for the first time, to present the potential for CPV discovery in a similar way to the MH case, i.e. quoting the power of the test as a function of exposure.

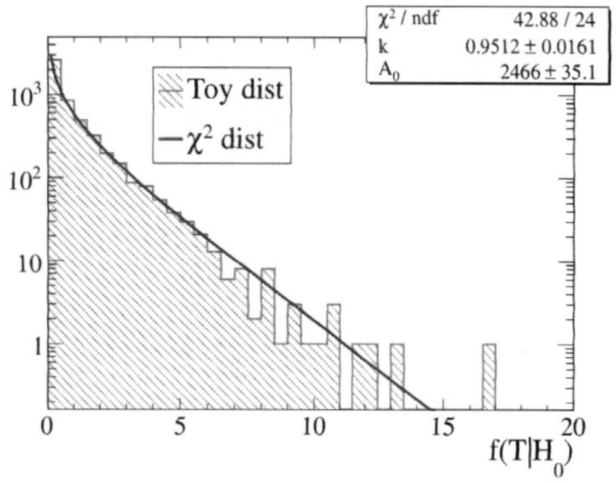

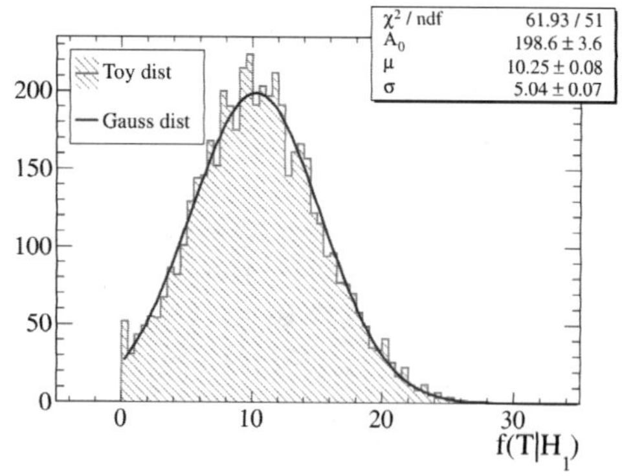

Figure 5. Distribution of the test statistic for the CPV test in the LBNO experiment, under the hypothesis of no CPV (left) and with $\delta_{CP} = \pi/2$ (right), with an exposure of $15{\times}10^{20}$ p.o.t.

Since the value of the test statistic is strongly dependent on δ_{CP}, as shown in Figure 6 (left) for the LBNO experiment, the power of the test will depend strongly on this parameter. The power of the LBNO experiment to exclude null CPV at two different CL is shown in Figure 6 (right) as a function of the exposure, for the two extreme cases $\delta_{CP} = \pi/2$ and $3\pi/2$, and it can be as low as zero for other values of the phase.

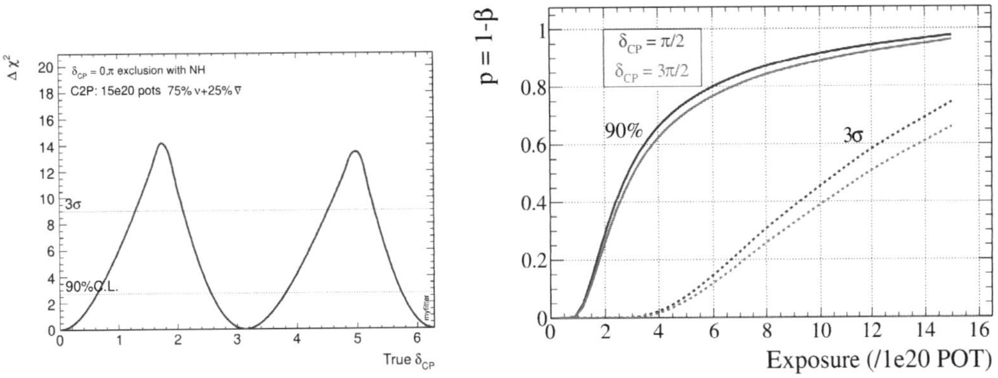

Figure 6. LBNO experiment. Left: average value of the test statistic for CPV test as a function of the δ_{CP} phase. Right: Power for exclusion of the non-existence of CPV as a function of exposure, for two CLs and for the two extreme values of δ_{CP}.

6. Conclusions

A review of some recent debate on how to evaluate and quote the sensitivity of future long-baseline neutrino experiment has been provided. The key conclusion is that, when dealing with problems of hypothesis testing, both CL and power of the test should be explicitly quoted.

The LBNO Collaboration has chosen to quote its sensitivity to Mass Hierarchy with a statistical power close to 1. A 5σ CL can be achieved by the experiment in at most 4-5 years of running, for either ordering.

A proposal to quote sensitivity to CP violation in a similar way has been shown.

A Bayesian approach to the Mass Hierarchy test has also been presented.

7. Acknowledgments

We are grateful to the European Commission for the financial support of the project through the FP7 Design Study LAGUNA-LBNO (Project Number 284518). We acknowledge the financial support of the UnivEarthS LabEx programme at Sorbonne Paris Cité (ANR-10-LABX-0023 and ANR-11-IDEX-0005-02).

References

[1] A. Stahl et al., Expression of Interest for a "Very long baseline neutrino oscillation experiment (LBNO)."; submitted to CERN SPSC, June, 2012. (CERN-SPSC-2012-021 (SPSC-EOI-007)).
[2] A. Rubbia, "Experiments for CP violation: A Giant liquid argon scintillation, Cerenkov and charge imaging experiment?", hep-ph/0402110; A. Rubbia et al., J. Phys. Conf. Ser. **171** (2009) 012020.
[3] X. Qian, A. Tan, W. Wang, J. J. Ling, R. D. McKeown and C. Zhang, Phys. Rev. D **86** (2012) 113011 [arXiv:1210.3651 [hep-ph]].
[4] E. Ciuffoli, J. Evslin and X. Zhang, JHEP **1401** (2014) 095 [arXiv:1305.5150 [hep-ph]].
[5] M. Blennow, P. Coloma, P. Huber and T. Schwetz, JHEP **1403** (2014) 028 [arXiv:1311.1822 [hep-ph]].
[6] M. Blennow, JHEP **1401** (2014) 139 [arXiv:1311.3183 [hep-ph]].
[7] S.K. Agarwalla *et al.* [LAGUNA-LBNO Collaboration], "The mass-hierarchy and CP-violation discovery reach of the LBNO long-baseline neutrino experiment," arXiv:1312.6520 [hep-ph].

[8] S. S. Wilks, The Annals of Mathematical Statistics, **9**, 1938.
[9] D. Franco, C. Jollet, A. Kouchner, V. Kulikovskiy, A. Meregaglia, S. Perasso, T. Pradier and A. Tonazzo *et al.*, JHEP **1304** (2013) 008 [arXiv:1301.4332 [hep-ex]].

Topical Research Meeting: Prospects in Neutrino Physics (NuPhys2013) IOP Publishing
Journal of Physics: Conference Series **598** (2015) 012006 doi:10.1088/1742-6596/598/1/012006

Short Baseline Neutrino Oscillation Experiments

Teppei Katori
Queen Mary University of London, London E1 4NS, UK
E-mail: t.katori@qmul.ac.uk

Abstract. Series of short baseline neutrino oscillation experiments provided unexpected results, and now they are called *short baseline anomalies*, and all indicates an existence of sterile neutrinos with a mass scale around 1 eV. The signals of short baseline anomalies are reported from 4 different classes of experiments. However, at this moment, there is no convincing theoretical model to explain such sterile neutrinos, and a single experiment to confirm 1 eV sterile neutrinos may be challenging. In this short note, we describe classes of short baseline neutrino oscillation experiments and their goals.

Classification of short baseline neutrino oscillation experiments

The short baseline anomalies come from 4 different classes of experiments [1]. Based on this, we can classify short baseline neutirno oscillation experiments into following 5 groups;

(i) test of LSND signal,
(ii) test of MiniBooNE signal,
(iii) test of reactor antineutrino anomaly,
(iv) test of Gallium anomaly, and
(v) others.

We discuss each of the above group in following sections. The last group is all other experiments, they are mainly experiments motivated by 1 eV sterile neutrinos and not short baseline anomalies themselves. Therefore many experiments in (v) are not short baseline oscillation experiments.

The short baseline anomalies are unsolved mysteries in this community, and they attract many theorists and experimentalists. The search of 1 eV sterile neutrino is one of the big branches of the neutrino experiment community [2]. Therefore, it is rather impossible to cover all experiments in this note, however, we try to cover most of experiments planned in the near future.

Journal of Physics: Conference Series **598** (2015) 012006 doi:10.1088/1742-6596/598/1/012006

1. LSND signal and experiments designed to test to it

LSND experiment

The origin of $\Delta m^2_{sterile} \sim 1$ eV^2 is the LSND experiment [3], where muon antineutrinos (0 to 53 MeV) are produced by pion decay-at-rest (DAR), and detected by a liquid scintillator detector at 31 m from the target. The LSND experiment measured $\bar{\nu}_e$ candidate events by utilizing the conincidence of the prompt Cherenkov radiation from the positron and the delayed neutron capture by a hydrogen.

$$\bar{\nu}_\mu \overset{oscillation}{\longrightarrow} \bar{\nu}_e + p \rightarrow e^+(Cherenkov) + n(capture) \; .$$

The LSND experiment observed an excess of $\bar{\nu}_e$ candidate events. This small ($< 1\%$ oscillation probability) but statistically significant signal is consistent with the presence of sterile neutrinos ($\bar{\nu}_\mu \rightarrow \nu_{sterile} \rightarrow \bar{\nu}_e$).

Meantime, the KARMEN experiment [4] excluded high Δm^2 region, and the Bugey experiment [5] excluded all low Δm^2 region of the LSND signal region in $\Delta m^2 - sin^2 2\theta$ plane. The combined result suggests the LSND signal is most likely due to sterile neutrinos around 1 eV region.

Experiments to test LSND signal

The LSND experiment has limited statistics, also, the duty cycle (nominal run, ~25 Hz) was high with a wide pulse. This allows LSND to accept large amount of cosmic backgrounds. Also, the detector was located to the direction of the primary beamline, and neutirnos from pion decay-in-flight (DIF) made additional backgrounds. Therefore, to test LSND signal, experiments are desired to have;

- LSND beam energy, baseline, and the detector which can tag $\bar{\nu}_e$ candidate events,
- higher statistics,
- known and narrow beam structure, and
- detector located on off-axis.

The promising experiments are off-axis liquid scintillator experiments at various neutron spallation sources in the world. The high pion-DAR $\bar{\nu}_\mu$ flux is available (and free!) from Oak Ridge National Laboratory (ORNL) [6], J-PARC Materials and Life Science Experimental Facility (MLF) [7], and European spallation source (ESS) [8]. On top of the high neutrino flux, proton pulses hitting the target to produce neutrons have well known beam structure. Therefore these experiments cover the desired features to test the LSND signal. Presently, OscSNS [6] is about to write a proposal, and J-PARC group [7] submitted proposal to J-PARC.

2. MiniBooNE signal and experiments designed to test it

MiniBooNE experiment

The MiniBooNE experiment is designed to test the LSND signal within the two massive neutrino oscillation hypothesis. However, muon (anti)neutrinos are now made by pion DIF at the Fermilab Booster neutrino beamline [9], and the baseline is 541 m from the target (pion decay length is ~18 m). The MiniBooNE detector is a spherical mineral oil based Cherenkov detector [10], and the $\nu_e(\bar{\nu}_e)$ candidate signals are measured as single isolated electron-like Cherenkov ring. In this way, the systematics of MiniBooNE is completely different from the LSND experiment, but MiniBooNE can test 1 eV sterile neutrino hypothesis, because of similar L/E with LSND.

$$\nu_\mu \stackrel{oscillation}{\longrightarrow} \nu_e + n \rightarrow p + e^+(Cherenkov) \ ,$$
$$\bar{\nu}_\mu \stackrel{oscillation}{\longrightarrow} \bar{\nu}_e + p \rightarrow n + \mathrm{e}^-(Cherenkov) \ .$$

The Booster neutrino beamline can run either in neutrino mode or in antineutrino mode, by focussing either positive or negative mesons. Since LSND signal was interpreted $\bar{\nu}_\mu \rightarrow \nu_{sterile} \rightarrow \bar{\nu}_e$ oscillations, running in antineutrino mode is more interesting. However, the antineutrino mode run suffers from lower statistics and higher backgrounds [11] (especially from the muon-neutrino contamination in the antineutrino mode beam [12]), therefore the experiment started in neutrino mode prior to the antineutrino mode running, and in the meantime systematics (the neutrino flux, neutrino interactions, and the detector response) were studied.

Unlike the LSND experiment, expected signal to noise is much lower. There are 2 dominant backgrounds of $\nu_e(\bar{\nu}_e)$ candidate events. The first one is the intrinsic $\nu_e(\bar{\nu}_e)$ contaminated in the beam. The majority of them are made by muon decay, therefore, MiniBooNE constrains them by simultaneously measuring ν_μ charged current quasi-elastic (CCQE) events [13, 14], where measured ν_μ is related to intrinsic ν_e through the pion decay chain ($\pi^+ \rightarrow \nu_\mu + \mu^+$, $\mu^+ -> \bar{\nu}_\mu + e^+ + \bar{\nu}_e$) in their simulation.

The second largest background is the misID of neutral current (NC) events, mainly NC π° production. Although a π° decays to two gamma rays which should be distinguishable from an electron (positron) Cherenkov ring, sometimes decay kinematics make two gamma rays look like one gamma ray (asymmetric decay, gamma rays are too close). Then, the Cherenkov ring from one gamma ray is indistinguishable from an electron (positron). For this, MiniBooNE internally measured NCπ° production, and the measured information was used to correct π° production rates in the simulation [15].

After the 10 years running in both neutrino and antineutrino mode, the MiniBooNE experiment observed excesses in both neutrino and antineutrino mode runs [16]. The final result is shown in Figure 1.

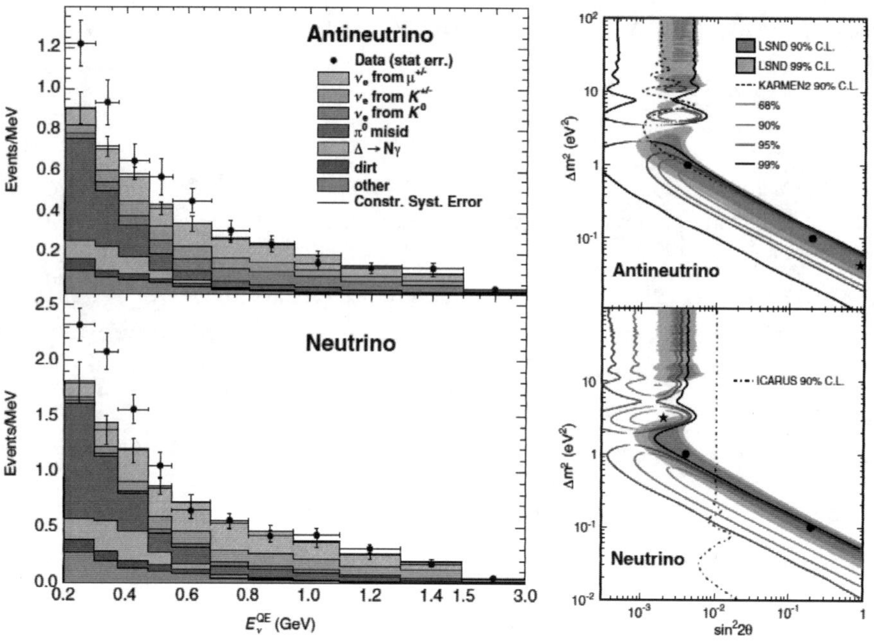

Figure 1. (color online) The final MiniBooNE oscillation results [16]. The top plots are antineutrino mode results, and bottom plots are neutirno mode results. The left plots show the reconstructed neutrino energy distribution of oscillation candidate events, and the right plots show the allowed region in Δm^2-$sin^2 2\theta$, where the best fit points are shown in black stars. Both modes show excesses in the low energy region, while the neutrino mode has higher statistical significance. On the other hand, the compatibility with the LSND signal is better in antineutrino mode.

Experiments to test MiniBooNE signal

The measured signal, especially in neutrino mode, does not quite agree with the expected sterile neutrino signal. The MiniBooNE detector cannot distinguish an electron (positron) and a gamma ray, therefore ν_μNC interaction with single gamma ray in the final state is a potential misID background. Therefore, to test the MiniBooNE signal, experiments are desired to have;

- MiniBooNE beam energy and baseline,
- ability to distinguish NC or CC interaction, or
- ability to distinguish an electron (positron) and a gamma ray.

The MiniBooNE+ was proposed to fulfill these criteria [17]. By doping scintillator (PPO) in the MiniBooNE detector mineral oil, MiniBooNE+ can measure scintillation light from the neutron capture. This allows statistical separation between ν_eCCQE interaction (higher proton multiplicity in the final state), and ν_μNC interactions (protons and neutrons are half-and-half in the final state). However, the proposal of MiniBooNE+ was not accepted by Fermilab recently.

The MicroBooNE experiment [18] is a new experiment on the Fermilab Booster neutrino beamline to test the MiniBooNE signal. It is also an important project for the future large liquid argon (LAr) TPC experiment, such as LBNE [19]. The MicroBooNE LArTPC detector has an ability to separate an electron from single gamma ray, by utilizing vertex-shower separation and dE/dx before developing the shower. This clearly tells if MiniBooNE excess is by an electron (positron) or a gamma ray [20]. The MicroBooNE experiment is under commissioning stage, and they expect beam data at the end of 2014.

Although the T2K experiment [21] is designed to measure the neutrino Standard Model (νSM) parameters and it does not use the Booster neutrino beam, J-PARC neutrino beam [22] has a similar beam peak (~600 MeV) as the Booster neutrino beam but is narrower, and the baseline to the near detector complex [23] is close (280 m) to what MiniBooNE has. Therefore, T2K is sensitive to the MiniBooNE signals ‡. The magnetic field in the near detector is a great advantage. It can allow the sign selection of the signal. The NC background (mostly ambient gamma rays) can be understood from the internal measurement [25].

3. Reactor antineutrino anomaly and experiments designed to test it

Reactor antineutrino anomaly

The re-evaluation of reactor electron antineutrino flux calculation provides consistently lower rate than world reactor data (about 6%) [26]. This, so called *reactor antineutrino anomaly* can be interpreted as neutrino oscillations with 1 eV sterile neutrinos [27].

Experiments to test Reactor anomaly

Reactor anomaly can be tested by small scale experiments, by measuring neutrino flux with small detectors with very short baseline (~15 m). To detect low energy reactor antineutrinos (~4 MeV), detector needs to be sensitive to low energy events. The common choice is the liquid scintillator detector, where large mass can be prepared at a relatively low cost.

There are number of such experiments designed for R&D of neutrino reactor monitoring for nuclear non-proliferation (SCRAAM [28], Nucifer [29], etc). These experiments are naturally served to test reactor antineutrino anomaly. Due to affordable cost of the experiments, several new experiments are also proposed to test the reactor antienturino anomaly (DANSS [30], PROSPECT [31], STEREO [32], etc).

‡ Current sterile neutrino search analysis in T2K is looking for ν_e disappearance [24], instead of ν_e appearance.

4. Gallium anomaly and experiments designed to test

Gallium anomaly

The two of pp-solar neutrino experiments, SAGE [33] and GALLEX [34], used highly radioactive sources to calibrate gallium detectors.

$$\nu_e +^{71} Ga \rightarrow^{71} Ge + e^- \ . \tag{1}$$

But some of these measurements using mega-curie ^{51}Cr or ^{37}Ar sources showed lower event rates than expected, and this so-called *Gallium anomaly* can be understood by neutrino oscilations with 1 eV sterile neutrinos [35].

Experiments to test Gallium anomaly

To test Gallium anomaly, a highly radio-active neutrino source and a very sensitive detector are required. Existing high sensitivity solar or reactor neutrino detectors (Borexino as "SOX" [36], KamLAND as "Ce-LAND" [37], SNO+ [39], DayaBay [38], etc) are good candidates for this purpose. Similarly, SAGE group proposed to build a new detector but reuse the liquid gallium from old detectors [40]. Proposed solar neutrino experiment (LENS [41], etc) and coherent scattering experiment (RICOCHET [42], etc) can look for 1 eV sterile neutrinos, too.

5. Others, 1 eV sterile neutrino searches

Tests by existing facilities

Once we assume 1 eV sterile neutrinos, some existing facilities are also sensitive to signals, even though the experiments are not originally designed to test 1 eV sterile neutrinos. The IceCube experiment [43] is designed to measure astrophysical ultra high energy neutrinos, however, 1 eV sterile neutrinos cause disappearance signals for >100 GeV high energy atmospheric neutrinos [44]. MINOS+ experiment [45] is an extension run of MINOS experiment [46, 47] during NOvA beam configuration era (medium energy NuMI, ~7 GeV peak at Sudan mine) [48]. One of physics goal of MINOS+ is to look for ν_μ disappearance signal due to 1 eV sterile neutrinos.

Tests by R&D facilities

Many R&D experiments for other purposes often look for 1 eV sterile neutrinos. For example IsoDAR experiment [49] look for sterile neutrino oscillation using ^{8}Li isotope made by the high power cyclotron. This cyclotron is a part of the R&D for the DAEδALUS experiment [50]. The νSTORM [51] experiment uses the muon storage ring, which is an important step for the future neutrino factory [52]. KDAR experiment [53] uses mono-energetic kaon DAR muon neutrinos (236 MeV), and the detector requires high resolution, such as LArTPC. This can be considered a part of LArTPC technology

LAr1-ND [54], and NESSiE [55], have detector R&D aspects for future large LArTPC experiments.

Ultimate 1 eV sterile neutrino search experiments

Experiments including precise measurement of oscillation probability with function of L/E can be considered in this group. LSND reloaded [56] was proposed to test short baseline neutrino oscillations by measuring oscillations with function of L/E in a large detector, such as gadolinium doped Super-Kamiokande. Similar concept may be applied to reactor antineutrino anomaly experiments and gallium anomaly experiments, where neutrino sources are small and low energy. In those experiments, precise L/E dependence measurement may be possible by either using a large detector or multiple small detectors, or moving sources and/or detectors. The precise oscillation measured in this way is a strong evidence of sterile neutrinos and it is a missing part from past experiments.

Aknowledgement

The author thanks Ranjan Dharmapalan for the careful reading of this manuscript. The author thanks to the organisers of "NuPhys2013, prospects in neutrino physics, (Institute of Physics, London, UK)" for the invitation to the conference.

References

[1] Abazajian K, Acero M, Agarwalla S, Aguilar-Arevalo A, Albright C *et al.* 2012 (*Preprint* 1204.5379)
[2] de Gouvea A *et al.* (Intensity Frontier Neutrino Working Group) 2013 (*Preprint* 1310.4340)
[3] Aguilar-Arevalo A *et al.* (LSND Collaboration) 2001 *Phys.Rev.* **D64** 112007 (*Preprint* hep-ex/0104049)
[4] Armbruster B *et al.* (KARMEN Collaboration) 2002 *Phys.Rev.* **D65** 112001 (*Preprint* hep-ex/0203021)
[5] Declais Y, Favier J, Metref A, Pessard H, Achkar B *et al.* 1995 *Nucl.Phys.* **B434** 503–534
[6] Allen R *et al.* (OscSNS Collaboration) 2013 (*Preprint* 1307.7097)
[7] Harada M *et al.* 2013 (*Preprint* 1310.1437)
[8] Baussan E *et al.* (ESSnuSB Collaboration) 2013 (*Preprint* 1309.7022)
[9] Aguilar-Arevalo A *et al.* (MiniBooNE Collaboration) 2009 *Phys.Rev.* **D79** 072002 (*Preprint* 0806.1449)
[10] Aguilar-Arevalo A *et al.* (MiniBooNE Collaboration) 2009 *Nucl.Instrum.Meth.* **A599** 28–46 (*Preprint* 0806.4201)
[11] Aguilar-Arevalo A *et al.* (MiniBooNE Collaboration) 2013 *Phys.Rev.* **D88** 032001 (*Preprint* 1301.7067)
[12] Aguilar-Arevalo A *et al.* (MiniBooNE Collaboration) 2011 *Phys.Rev.* **D84** 072005 (*Preprint* 1102.1964)
[13] Aguilar-Arevalo A *et al.* (MiniBooNE Collaboration) 2008 *Phys.Rev.Lett.* **100** 032301 (*Preprint* 0706.0926)
[14] Aguilar-Arevalo A *et al.* (MiniBooNE Collaboration) 2010 *Phys.Rev.* **D81** 092005 (*Preprint* 1002.2680)

[15] Aguilar-Arevalo A *et al.* (MiniBooNE Collaboration) 2008 *Phys.Lett.* **B664** 41–46 (*Preprint* 0803.3423)
[16] Aguilar-Arevalo A *et al.* (MiniBooNE Collaboration) 2013 *Phys.Rev.Lett.* **110** 161801 (*Preprint* 1207.4809)
[17] Dharmapalan R *et al.* (MiniBooNE+ Collaboration) 2013 (*Preprint* 1310.0076)
[18] Camilleri L (MicroBooNE Collaboration) 2013 *Nucl.Phys.Proc.Suppl.* **237-238** 181–183
[19] Akiri T *et al.* (LBNE Collaboration) 2011 (*Preprint* 1110.6249)
[20] Fleming B T 2011 *J.Phys.Conf.Ser.* **308** 012007
[21] Abe K *et al.* (T2K Collaboration) 2014 *Phys.Rev.Lett.* **112** 061802 (*Preprint* 1311.4750)
[22] Abe K *et al.* (T2K Collaboration) 2013 *Phys.Rev.* **D88** 032002 (*Preprint* 1304.0841)
[23] Abe K *et al.* (T2K Collaboration) 2011 *Nucl.Instrum.Meth.* **A659** 106–135 (*Preprint* 1106.1238)
[24] SGALABERNA D 2014 *PoS* **EPS-HEP2013** 011
[25] Abe K *et al.* (T2K Collaboration) 2014 (*Preprint* 1403.2552)
[26] Mueller T, Lhuillier D, Fallot M, Letourneau A, Cormon S *et al.* 2011 *Phys.Rev.* **C83** 054615 (*Preprint* 1101.2663)
[27] Mention G, Fechner M, Lasserre T, Mueller T, Lhuillier D *et al.* 2011 *Phys.Rev.* **D83** 073006 (*Preprint* 1101.2755)
[28] Bowden N, Bernstein A, Dazeley S, Svoboda R, Misner A *et al.* 2008 *J.Appl.Phys.* (*Preprint* 0808.0698)
[29] Porta A (Nucifer Collaboration) 2010 *Journal of Physics: Conference Series* **203** 012092
[30] Danilov M (DANSS Collaboration) 2013 (*Preprint* 1311.2777)
[31] Djurcic Z, Hans S, Yeh M, Blucher E, Johnson R *et al.* 2013 (*Preprint* 1309.7647)
[32] LHUILLIER D 2014 *PoS* **EPS-HEP2013** 522
[33] Abdurashitov J *et al.* (SAGE Collaboration) 2009 *Phys.Rev.* **C80** 015807 (*Preprint* 0901.2200)
[34] Hampel W *et al.* (GALLEX Collaboration) 1999 *Phys.Lett.* **B447** 127–133
[35] Giunti C and Laveder M 2011 *Phys.Rev.* **C83** 065504 (*Preprint* 1006.3244)
[36] Bellini G *et al.* (Borexino Collaboration) 2013 (*Preprint* 1304.7721)
[37] Gando A, Gando Y, Hayashida S, Ikeda H, Inoue K *et al.* 2013 (*Preprint* 1309.6805)
[38] Dwyer D, Heeger K, Littlejohn B and Vogel P 2013 *Phys.Rev.* **D87** 093002 (*Preprint* 1109.6036)
[39] O'Keeffe H, Kraus C, Wan Chan Tseung H and Wilson J (SNO+ Collaboration) 2012 *Nucl.Phys.Proc.Suppl.* **229-232** 552
[40] Gorbachev V, Veretenkin E, Gavrin V, Dan'shin S, Ibragimova T *et al.* 2013 *Phys.Atom.Nucl.* **76** 1507–1511
[41] Grieb C, Link J and Raghavan R 2007 *Phys.Rev.* **D75** 093006 (*Preprint* hep-ph/0611178)
[42] Anderson A, Conrad J, Figueroa-Feliciano E, Scholberg K and Spitz J 2011 *Phys.Rev.* **D84** 013008 (*Preprint* 1103.4894)
[43] Aartsen M *et al.* (IceCube) 2013 *Science* **342** 1242856 (*Preprint* 1311.5238)
[44] Barger V, Gao Y and Marfatia D 2012 *Phys.Rev.* **D85** 011302 (*Preprint* 1109.5748)
[45] Evans J (MINOS) 2013 *Adv.High Energy Phys.* **2013** 182537 (*Preprint* 1307.0721)
[46] Adamson P *et al.* (MINOS Collaboration) 2013 *Phys.Rev.Lett.* **110** 251801 (*Preprint* 1304.6335)
[47] Adamson P *et al.* (MINOS Collaboration) 2011 *Phys.Rev.Lett.* **107** 011802 (*Preprint* 1104.3922)
[48] Messier M (NOvA Collaboration) 2013 (*Preprint* 1308.0106)
[49] Bungau A, Adelmann A, Alonso J, Barletta W, Barlow R *et al.* 2012 *Phys.Rev.Lett.* **109** 141802 (*Preprint* 1205.4419)
[50] Aberle C, Adelmann A, Alonso J, Barletta W, Barlow R *et al.* 2013 (*Preprint* 1307.2949)
[51] Adey D *et al.* (nuSTORM Collaboration) 2014 *Phys.Rev.* **D89** 071301 (*Preprint* 1402.5250)
[52] Tunnell C D, Cobb J H and Bross A D 2011 (*Preprint* 1111.6550)
[53] Spitz J 2012 *Phys.Rev.* **D85** 093020 (*Preprint* 1203.6050)
[54] Adams C *et al.* 2013 (*Preprint* 1309.7987)
[55] Antonello M, Bagliani D, Baibussinov B, Bilokon H, Boffelli F *et al.* 2012 (*Preprint* 1203.3432)
[56] Agarwalla S K and Huber P 2011 *Phys.Lett.* **B696** 359–361 (*Preprint* 1007.3228)

Solar neutrinos: experimental review and prospectives

Marco Pallavicini
Dipartimento di Fisica Università di Genova and Sezione INFN di Genova
E-mail: marco.pallavicini@ge.infn.it

Abstract. Solar neutrino physics, after the golden age which led to the discovery of neutrino oscillations, is about to enter the precision era. We summarise here the current experimental knowledge in the field and the future perspectives, showing that solar neutrino detectors are and will remain in the near future a fundamental tool for a deeper understanding of stars, neutrinos, and fundamental physics.

1. Historical introduction

The historical development of solar neutrino physics can be divided into two periods. The first, which lasted from the early 70s until 2001, began with the pioneering work of the radiochemical ^{37}Cl experiment at Homestake (1970-1994, [1]), grew with the study of high energy (E>6 MeV) solar neutrinos by means of the water Cherenkov detectors at Kamioka mine in Japan (KamiokaNDE 1983-1995 [2][3] and Super-KamiokaNDE 1996-present [4]), continued with the detection of low energy neutrinos with gallium radiochemical experiments (Gallex [5], SAGE [6]), and was concluded by the discovery of solar neutrino oscillations by SNO [7][8].

The second period started after the establishment of neutrino oscillations and the beginning of the high precision era of solar physics, which has been initiated mainly by the SNO, KamLAND and Borexino experiments and which continues to develop thanks to upgrades and improvements of existing experiments and the design of future projects.

The first period, certainly a golden age for neutrino physics, has witness the first detection of solar neutrinos at different energies, which have proved unambiguously that the Sun's energy has a nuclear origin, the development of a very refined Standard Solar Model (SSM), the recognition of a sharp discrepancy between the model itself and the observed neutrino rates (the so called Solar Neutrino Problem, SNP), and the triumphal discovery of neutrino flavour oscillations as the sole possible explanation of that discrepancy.

Although solar neutrino oscillations were, since the very beginning, a possible explanation of the SNP, the favour increased only when a clear evidence of oscillations was found in 1998 by means of atmospheric neutrinos [9].

The observed deficit is explained by the fact that most solar neutrino experiments are either insensitive to muon or tau neutrinos, or have a suppressed cross section. The fact that the deficit is energy dependent requires the so called MSW effect [10][11][12][13], the resonant enhancement of flavour conversion due to the coherent interaction of neutrinos in the solar medium. However, it was the SNO experiment that closed the story. The use of heavy water made the experiment able to measure separately the electron neutrino and the total neutrino flux at Earth, proving without ambiguities that the total neutrino flux at is the same as predicted by the SSM and that the electron neutrino flux is reduced by oscillations. Later measurements with enhanced neutron detection capability, first by dissolving salt in water and later with ^{3}He detectors, have confirmed the result. The oscillation picture was confirmed soon by the KamLAND reactor experiment [14].

The second period, which started after the discovery of neutrino oscillations, is continuing nowadays. Although main goal of solar neutrino physics has slightly changed, the focus is still both on fundamental neutrino physics and on solar science. In this paper we summarise the results obtained since 2001 on solar neutrinos by many different experiments, and particularly by Borexino at the Laboratory Nazionali del Gran Sasso. At the time of writing, the running solar neutrino experiments are Borexino, Sage, Super-kamiokaNDE and KamLAND. The SNO+ collaboration is upgrading the detector to start a new solar neutrino phase with liquid scintillator instead of heavy water. Other experiments which are under discussion include solar neutrino programs. For a recent review of solar neutrino physics see e.g. [15].

2. Solar physics AD 2013

For thirty years solar neutrino physics was driven by the solar neutrino problem, which was finally understood in terms of the Pontecorvo-Maki-Nakagawa-Sakata (PMNS) neutrino flavour mixing matrix, the analogous of the CKM matrix in the quark sector. Now that neutrino oscillations have been established and all mixing angles, including the relatively small θ_{13}, have been measured, precise spectroscopy of the full spectrum of solar neutrinos is possible.

This fact offers two important scientific opportunities: on one hand, with weak interaction uncertainties removed, solar neutrinos can be used to probe possible limitations of the Standard Solar Model, such as, in particular, the uncertainties in the metal content (metallicity problem), and the effect of rotation, mixing, or magnetic fields. Future measurements at higher precision with respect to the one currently available, might also allow to compare the neutrino luminosity with the photon one, and study the thermal equilibrium of the Sun in a unique way. On the other hand, the neutrinos coming from the Sun remain important to fundamental physics: the spectral shapes and fluxes of the various sources are known rather precisely, offering the possibility to study neutrino properties such as, e.g. neutrino magnetic moments, the existence of non-standard neutrino-electron interactions, or the existence of sterile neutrinos.

Some of these measurement are possible within a relatively short time scale, others are more for the future. In particular, a more precise measurement of ^{7}Be, pep, and ^{8}B neutrinos might constraint the P_{ee} to a level which might be sufficient to probe the MSW effect and search for new physics. A first direct measurement of pp neutrinos might be behind the corner (Borexino), although probably not with sufficient precision such as it would be needed to probe the solar luminosity constraint.

The metallicity problem might be solved by a precise direct measurement of CNO neutrinos. This is certainly a very difficult task which might be faced wither by Borexino or by the SNO+ experiment. A future detector such as LENS might measure all neutrino fluxes neatly and with small background.

The paper is structured as follows; section 2 summarise the many important goals of solar neutrino physics today; section 3 is devoted to the results obtained on ^{8}B neutrinos by Borexino, by KamLAND, and by the SNO and Super-KamiokaNDE experiments, which have made continuos improvement on their detectors and on their analysis methods, reducing the energy threshold and improving the statistics ; section 4 is devoted to a summary of the results obtained on sub-MeV neutrinos by Borexino, definitely the main solar neutrino results obtained recently; section 5 outlines the main future prospectives, both of existing experiments and of new projects under construction or foreseen for the near future; finally, in section 6 I will discuss briefly the use of the Borexino detector as a tool to search for sterile neutrinos by means of strong neutrino

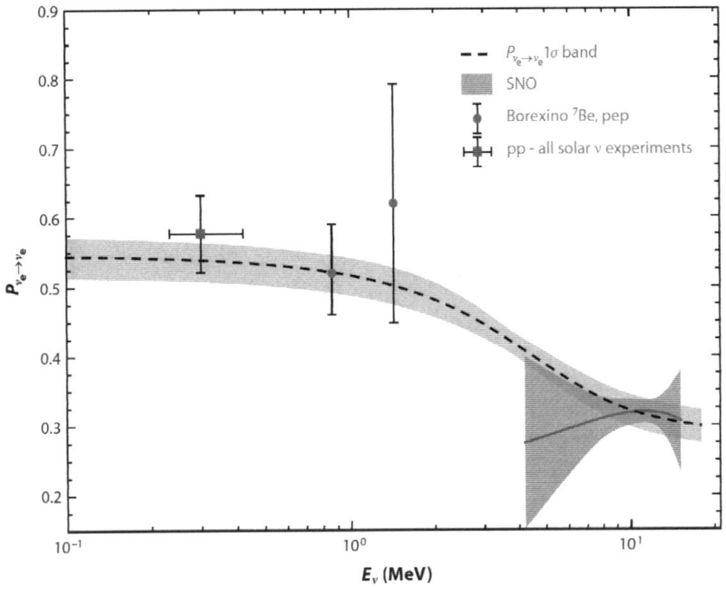

Figure 1. Theoretical expectation for the electron neutrino survival probability P_{ee} according to the LMA-MSW solution (dashed line, with grey band representing theoretical uncertainty) and current experimental knowledge from all solar neutrino experiment. The low energy point is calculated from gallium solar neutrino data using measured ^{7}Be data from Borexino. Figure taken from [15].

and anti–neutrino artificial sources.

3. ^{8}B neutrinos: up-turn or no up-turn ?

A very interesting feature of the MSW-LMA solution is the so called "up-turn", i.e. the rise of the P_{ee} right below the ^{8}B solar neutrino energies. It is important for two main reasons: on one hand, it is the signature of the MSW effect, being the solar neutrino oscillations essentially vacuum oscillations below the critical energy $E_\nu \approx$1.9 MeV ($P_{ee} \simeq$0.57 for $E_\nu \ll$ 1.9MeV) and being dominated by matter effects at high energy ($P_{ee} \simeq$0.31 for $E_\nu \gg$ 1.9MeV).The region around 1.9 MeV is called the transition region and the rise of the P_{ee} below $\approx$3 MeV the up-turn (see Fig. 1). On the other hand, significant deviations from the expected P_{ee} shape might provide hints of new physics effects, such as, e.g., non-standard neutrino interactions in the Sun or in the detector, or the existence of sterile neutrinos.

Four different experiments (SK, SNO, Borexino and KamLAND) using two different techniques have measured the solar neutrino flux in the range 3-10 MeV, where the spectrum is totally dominated by ^{8}B neutrinos. Fig. 2 summarises the result of the experiments.

Borexino detects solar neutrinos by means of elastic scattering of electrons in a liquid scintillator and has been the first experiment capable of measuring ^{8}B events with an electron recoil energy threshold at 3 MeV. The top-left plot in Fig. 2 shows the Borexino result, which is well consistent with LMA-MSW prediction, except perhaps for the lowest energy bin. KamLAND (top-right plot in Fig. 2), which uses the same technique with a higher energy threshold, finds a similar result, again with an apparent deficit in the lowest energy bin.

SNO has reduced the energy threshold with a dedicated analysis effort called LETA [18]. Its result (bottom-left plot in Fig. 2) is well consistent with LMA-MSW except two points at low energy, which yield a result lower than expected.

Super-kamiokaNDE measures the ^{8}B neutrino flux by detecting Cherenkov light of scattered electrons. A major improvement of electronics and triggering capabilities (SK-IV phase, since 2008) has lowered the detection threshold of solar neutrinos down to 3.5 MeV. This fact has opened the possibility to study the "up-turn" region. The bottom-right plot in Fig. 2 (from [19]) shows that LMA-MSW solutions fits very well the energy distribution of the events, with a statistically not significant tension in the very low energy bins.

Although not statistically significant, all experiments find consistently lower signal in the low energy portion of the ^{8}B solar neutrino spectrum, where the "up-turn" of the P_{ee} is expected. It is too early to say whether these are just statistical fluctuations, systematic errors close to the detection thresholds, or hints of new physics. We just underline that the latter possibility is of course very interesting and motivates the efforts of existing or future experiments. SK and Borexino will continue to take data for some years and will improve the result.

Recently, the SK collaboration has published a very interesting measurement of ^{8}B neutrino flux day–night asymmetry [20]. At ^{8}B energies the regeneration of the neutrinos during their path across the Earth is not negligible and the cited paper has provided the first evidence of that effect, although only at 2.7σ level. If confirmed, this result would be the first direct detection of MSW effect of neutrinos in matter.

4. sub-MeV solar neutrino detection

Borexino is the only experiment that has a sufficiently low detection threshold and a low enough intrinsic radioactive background to detect sub-MeV solar neutrinos. This section is therefore limited to Borexino results.

The main goal of the experiment was the detection of the monochromatic neutrinos that are emitted in the electron capture of ^{7}Be in the Sun with 5% precision [21][22]. In order to achieve such a goal the intrinsic contamination of the liquid scintillator and the external radiation must be reduced to unprecedented low levels. In particular, design goals of Borexino required ^{238}U and ^{232}Th contents below 10^{-16} g/g, $^{40}K_{nat}$ content below 10^{-14} g/g and a few counts per day maximum in the fiducial volume from ^{85}Kr and ^{39}Ar. These goals, thanks to a long and dedicated effort and the development of innovative purification techniques, have all been met and in some cases by far exceeded, as e.g. for ^{238}U and ^{232}Th contents, which are both below 10^{-18} g/g. See [23], [24], [25], [26] and references therein for more details.

The precision measurement of ^{7}Be has been obtained during the first stage of the experiment [27][28]. Real time detection of the events and detector stability allowed also to study the day-night effect of the ^{7}Be solar neutrino signal, which yielded the exclusion of the LOW solution of the neutrino oscillation based on solar data alone [29].

Finally, the very low background of the detector, a refined analysis on threefold coincidences and an innovative discrimination method of β^+ events based on the positronium formation, made it possible to explore the 1-2 MeV region with unprecedented sensitivity. This led to the first observation of solar neutrinos from the basic pep reaction [30] and to the best upper limit for the CNO production in the Sun. Fig. 1 summarises well the impact of Borexino measurements on our current knowledge of P_{ee}. Work is in progress to achieve a direct measurement of pp-neutrinos flux. A direct measurement of CNO neutrinos is also among the goals of the experiment, although detector stability limitations might prevent a reliable measurement.

5. Future perspectives

Several experiments are foreseen for the forthcoming years. It is not possible, within the allowed space of this paper, to give a comprehensive review of all proposed projects. I will limit the discussion for those who are already approved or very mature for approval.

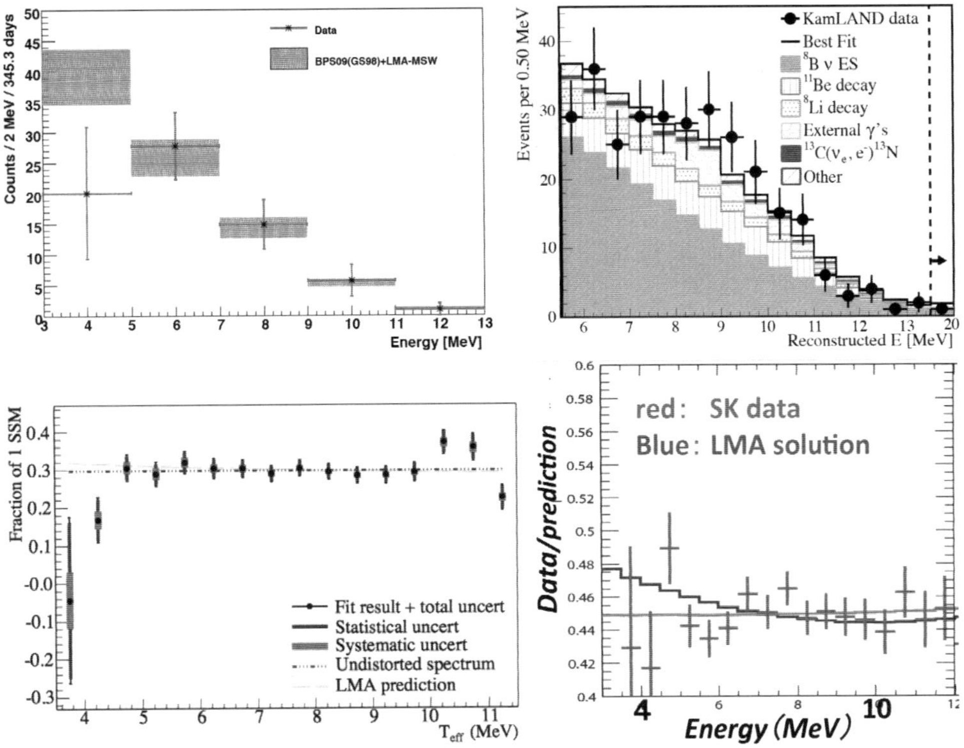

Figure 2. Results for the search of the ^{8}B up-turn from four experiments: Top Left) Borexino [16]; Top Right) KamLAND [17]; Bottom Left) SNO [18]; Bottom Right) Super-kamiokaNDE [19]. Although not statistically significant, all experiments find consistently lower signal in the low energy portion of the ^{8}B solar neutrino spectrum, where the "up-turn" of the P_{ee} is expected.

Among the running experiments, Borexino and Super-kamiokaNDE are the ones that could yield the most interesting results.

After a successful purification campaign, the Borexino experiment has reduced the scintillator background even further with respect to 2007-2010 data. In particular, the ^{85} background was significantly reduced, opening the way to a first direct measurement of pp neutrinos. Borexino might possibly be able to put a strong constraint on CNO solar neutrinos, which would be very useful to solve the metallicity saga. However, besides a strong reduction of ^{210}Bi background, which was partially achieved by means of purifications, a much higher detector stability is required to be able to measure such a low signal (about 3 times smaller than the already difficult pep signal). Further purification campaigns might be attempted 2015, possibly with an improved water plant. If successful, a lower ^{210}Bi background united with an improved detector stability might open a window of opportunity for CNO neutrinos. *Se son rose, fioriranno.*

The most important future project in solar neutrino physics, besides those already running, is SNO+ [31].

The SNO+ collaboration is constructing a liquid scintillator experiment re-using the vessel and the photomultipliers of the SNO experiment. The scintillator is linear alkylbenzene (LAB) doped of PPO, which is safe (high flash point), low cost, and compatible with the existing acrylic vessel. The project has two main goals: the study of pep and CNO neutrinos, and the search of neutrino less double beta decay with Te dissolved in the scintillator.

As far as solar neutrinos are concerned, the higher depth of the Sudbury mine compared to Gran Sasso and Kamioka yields a much lower cosmogenic background from ^{11}C (700 times less than Kamioka, 100 times less than LNGS). According to the simulations, SNO+ may be able to measure the CNO neutrino flux with 6% error. This precision is sufficient to measure the metal content of the solar core in a model independent way and possibly solve the solar metallicity problem. The expected signal in SNO+ for pep and CNO neutrinos is shown in Fig. 3. This sensitivity, however, will be possible only the the background induced by ^{210}Bi is at least as low as the Borexino one, or maybe better.

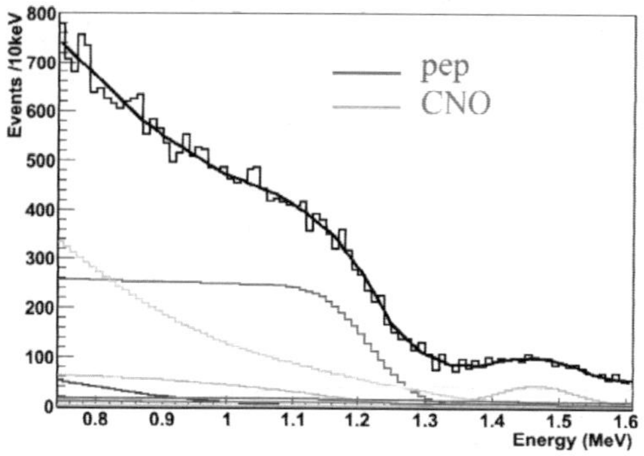

Figure 3. The signal expected in SNO+ in three years of data taking for CNO and pep neutrinos. The low cosmogenic ^{11}C background and the high mass make SNO+ may allow a 6% measurement of CNO neutrinos in three years of data.

Another more ambitious project for solar neutrino physics is LENS, Low Energy Neutrino Spectroscopy. This experiment aims at the precise measurement of all solar neutrino components by detecting neutrinos via charged current interactions (inverse beta decay) with ^{115}In.

The experimental tool used in the LENS detector for the detection of solar neutrinos is the tagged capture of ν_es on ^{115}In via charged current inverse beta decay. The tagged technique has two outstanding advantages over competing experiments: first, there is a one-to-one correspondence between the incoming neutrino energy and the measurable electron energy $E_\nu = E_e + Q_d$ (Q_d: capture threshold), and second, the γ-cascade allows the application of time/space coincidence techniques to suppress ubiquitous radioactive backgrounds as well as the inherent background from the beta decay of ^{115}In.

A modular detector is required. The LENS detector is a novel Scintillation Lattice Chamber, an optically segmented, three dimensional array of 0.5 l cells of liquid scintillator loaded with 8-10 wt% Indium. The scintillation signal from each cell will be always viewed by the same set of 3-6 phototubes. Thus, the full scale LENS of 125 tons InLS, though large in size, is in essence, a large array of small detectors capable of bench-top precision nuclear spectroscopy. It will provide extraordinary spatial resolution in a large mass of liquid scintillator through segmentation rather then time of flight information, which allows adequate background rejection using the time/space coincidence tag even for low energy (100 keV) events.

If successful, this experiment might really represent the future of solar neutrino physics because of the superbe energy resolution possible for all neutrino energies. A spectrum of the expected signal in LENS is shown in Fig. 4 together with a conceptual design of the detector.

Finally, several new projects include solar neutrino in their scientific program. Among those, we briefly mention Clean [33] and Xmass [34]. We refer to the references for more details.

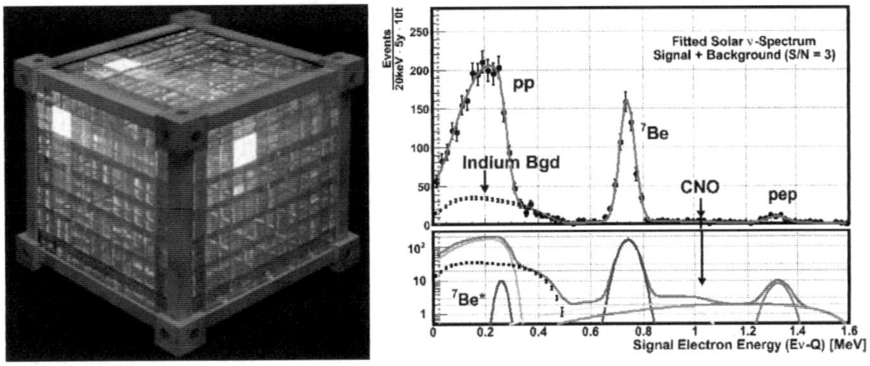

Figure 4. Left) Lens conceptual design. Isotropically emitted scintillation light is channeled to the outside of the detector along the main axes via total internal reflection Right) Lens expected spectral resolution on solar neutrinos. If the detector will work as anticipated, LENS will give a beautiful spectral measurement of all solar neutrino components separately, with very high precision and low background.

All these programs want to use the nice features of cryogenic liquid as active targets for solar neutrinos and search for neutrino-less double beta decay and dark matter. Particularly, liquid noble gasses have a good light yield, high density and can be purified thoroughly, yielding good signals and low background. Large mass detectors of this type maybe a crucial tool for next generation solar neutrino experiments.

6. The Borexino detector as a tool to search for sterile neutrinos

The superbe radio purity of the detector, its large size and its capability to detect and locate both neutrino and anti–neutrino events, make Borexino an ideal instrument for the study of short distance neutrino oscillations by means of intense neutrino sources located close to or, maybe in the future, inside the detector. The SOX project [39] aims at the search of sterile

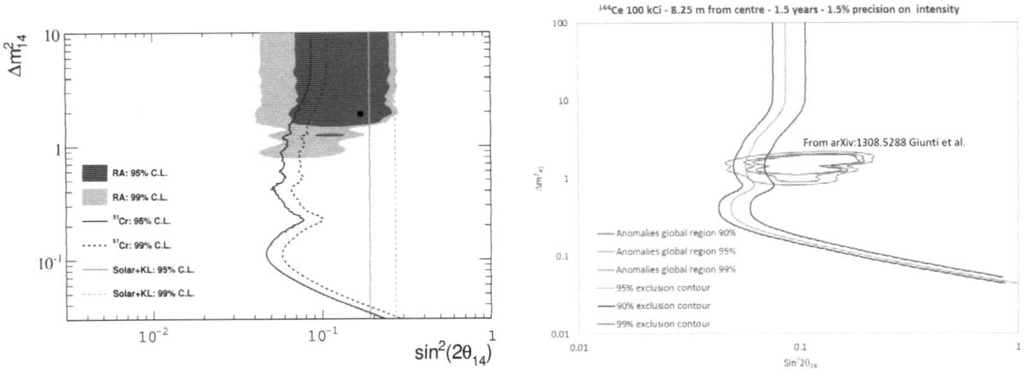

Figure 5. Left) ^{51}Cr SOX sensitivity compared with the allowed reactor anomaly region. The yellow line is the limit from solar neutrino analysis. Right) ^{144}Ce SOX sensitivity compared with the region obtained as global fit from all neutrino anomalies [37].

neutrinos or the study of other non-standard short distance effects using an intense 5-10 MCi neutrino source made with about 35 kg of ^{51}Cr activated in a nuclear reactor and with a 100-150 kCi anti-neutrino source made of ^{144}Ce [38]. The experiment will probe the so called reactor and gallium anomalies and search for sterile neutrinos with high sensitivity. Fig. 5 shows the sensitivity of the experiment both to neutrino (left plot) and anti-neutrino sources (right plot). The relatively large size of the Borexino fiducial volume (about 6 m diameter for neutrino events, about 8 m for anti-neutrino events) allow to detect directly oscillations waves if the δm_s^2 with the sterile component is of the order of 1 eV^2. This makes SOX a rather unique experiment. See [39] for more details.

Acknowledgements
I wish to thank the Borexino Collaboration for the beautiful work we have done together in these years, and the Laboratori Nazionali del Gran Sasso for their continual support. My special thanks also to the organisers of this conference for the kind and efficient hospitality. SOX is funded by ERC-2012-AdG project N. 320873 (FP7), P.I Marco Pallavicini.

References

[1] B. T. Cleveland et al., Astroph. J. , **496** (1998) 505-526
[2] K.S. Hirata et al. (KamiokaNDE-II collaboration), Phys. Rev. Lett **63** (1989) 16-19
[3] Y. Fukuda et al. (KamiokaNDE-III collaboration) Phys. Rev. Lett. **77** (1996) 1683-1686
[4] S. Fukuda et al. (Super-KamiokaNDE-I collaboration) Phys. Lett. **B539** (2002) 179-187
[5] P. Anselmann et al. (Gallex collaboration) Phys. Lett. **B285** (1992) 376-389
[6] J. N. Abdurashitov et al. (SAGE collaboration), Phys. Lett. **B328** (1994) 234-248
[7] Q. R. Ahmad et al. (SNO collaboration), Phys. Rev. Lett. **87** (2001) 071301
[8] Q. R. Ahmad et al. (SNO collaboration), Phys. Rev. Lett. **89** (2002) 011301
[9] Y. Fukuda et al. (Super-KamiokaNDE-I collaboration) Phys. Rev. Lett. **81** (1998) 1562-1567;
[10] L. Wolfenstein, Phys. Rev. **D17** (1978) 2369
[11] L. Wolfenstein, Phys. Rev. **D20** (1978) 2634
[12] S.P. Mikheyev and A.Y. Smirnov Sov. J. Nucl. Phys. **42** (1985) 1913
[13] S.P. Mikheyev and A.Y. Smirnov Nuovo Cim. **9C** (1986) 17
[14] K. Eguchi et al. (KamLAND collaboration) Phys. Rev. Lett. **90** (2003) 021802
[15] W.C. Haxton, R.G. Hamish Robertson and A.M. Serenelli, Annu. Rev. Astron. Astrophys. 51 (2013) 21-61
[16] G. Bellini et al. (Borexino Collaboration) Phys. Rev. **D82** (2010) 033006
[17] B. Aharmim et al. (KamLAND Collaboration) Phys. Rev. **C84** (2011) 035804
[18] B. Aharmim et al. (SNO Collaboration) Phys. Rev. **C81** (2010) 055504
[19] Y. Suzuki (for the Super-kamiokaNDE Collaboration) Neutrino Telescopes (2013)
[20] A. Renshaw et al. (Super-kamiokaNDE collaboration) Phys. Rev. Lett. **112** (2014) 091805
[21] G. Alimonti *et al.* (Borexino Collaboration), Astropart. Phys. **16** (2002) 205
[22] C. Arpesella *et al.* (Borexino Collaboration), Astropart. Phys. **18** (2002) 1
[23] H. O. Back et al. (Borexino Collaboration), Nucl. Instrum. Methods A **609** (2009) 58
[24] H. O. Back et al. (Borexino Collaboration), Phys. Lett. B 563 (2003) 23
[25] H. O. Back et al. (Borexino Collaboration), Phys. Lett. B 525 (2002) 29
[26] G. Bellini et al. (Borexino Collaboration) arXiV 1308.0443v1, submitted to Phys. Rev. D
[27] C. Arpesella *et al.* (Borexino Collaboration), Phys. Lett. B **658** (2008) 101
[28] G. Bellini et al. (Borexino Collaboration) Phys. Rev. Lett. **107** (2011) 141302
[29] G. Bellini et al. (Borexino Collaboration) Phys. Lett. **B707** (2012) 22
[30] G. Bellini et al. (Borexino Collaboration) Phys. Rev. Lett. **108** (2012) 051302
[31] N. Tolich for the SNO+ Collaboration NNN 09, http://nnn09.colostate.edu/Talks/Session05/
[32] C. Grieb, Nucl. Phys. **B 168** (2007) 122
[33] D. McKinsey et al., Astropart.Phys. **22** (2005) 355-368
[34] K. Abe for Xmass Collaboration, Jour. of Phys. **120** 4 (2008)
[35] M. Cribier et al., Phys. Rev. Lett. **107** (2011) 201801
[36] G. Bellini et al., (Borexino Collaboration) JHEP **08** (2013) 038
[37] C. Giunti et al. Phys. Rev. **D88** (2013) 073008
[38] M. Cribier et al., Phys. Rev. Lett. **107** (2011) 201801
[39] G. Bellini et al. (SOX collaboration) JHEP **08** (2013) 038

Neutrinoless double beta decay experiments

M C Chen

Department of Physics, Queen's University, Kingston, Ontario K7L 3N6, Canada

E-mail: mchen@queensu.ca

Abstract. Understanding the charge conjugation nature and the absolute mass scale of neutrinos will lead to valuable insights into physics beyond the Standard Model. Both of these aspects can be probed by the observation of neutrinoless double beta decay. The experimental search for neutrinoless double beta decay is thus an important, next step in neutrino physics. Several experiments have recently started taking data and several more are under construction. The sensitivity of upcoming double beta decay experiments will begin to probe neutrino masses as would be suggested by the inverted neutrino mass hierarchy. A summary of the capabilities and status of several double beta decay experiments was presented at NuPhys 2013.

1. Introduction

Double beta decay is the decay of a nucleus in which two neutrons change into two protons, with two outgoing electrons and two antineutrinos being emitted. This process was first proposed by Goeppert-Meyer in 1935. This second-order weak decay, which will be referred to as $2\nu\beta\beta$ decay in this summary, is quite rare involving half-lives of the order 10^{20} years or more. After Majorana proposed in 1937 that neutrinos could be their own antiparticles (we now call particles with this charge-conjugation nature "Majorana" fermions, as opposed to "Dirac" fermions), Furry pointed out in 1939 that double beta decay might proceed without neutrino emission. As illustrated in Fig. 1, the process essentially involves one of the antineutrinos emitted in the decay being absorbed as a neutrino by the other transforming nucleon. It is clear that neutrinoless double beta decay ($0\nu\beta\beta$ decay) can occur if and only if neutrinos are Majorana particles.

One observes from the diagram for the process (and an understanding of the weak interaction) that the helicity of the neutrino has to "flip" in order for it to be absorbed. Consequently, the transition amplitude for $0\nu\beta\beta$ must be related to the wrong-handed component of the neutrino propagator, and thus proportional to m_ν/E_ν. More precisely, the amplitude for this process involves all three neutrino masses and the PMNS mixing matrix components U_{ei} squared, since there are two vertices involving the neutrino propagator and electrons in Fig. 1. We define:

$$|\sum_i m_i U_{ei}^2| \equiv \langle m_{\beta\beta}\rangle \qquad (1)$$

where the value $\langle m_{\beta\beta}\rangle$ is the known as the effective Majorana neutrino mass for double beta decay. As the decay rate for neutrinoless double beta decay depends on the amplitude squared, it will be proportional to $\langle m_{\beta\beta}\rangle^2$. It is important to note that the amplitude is the coherent sum of complex-valued U_{ei} elements of the PMNS matrix, including the so-called "Majorana phases." The consequence of this summation is the possibility of cancellations in the amplitude; Fig. 2 from [1] illustrates this possible effect in the allowed regions that dip down to small values. In

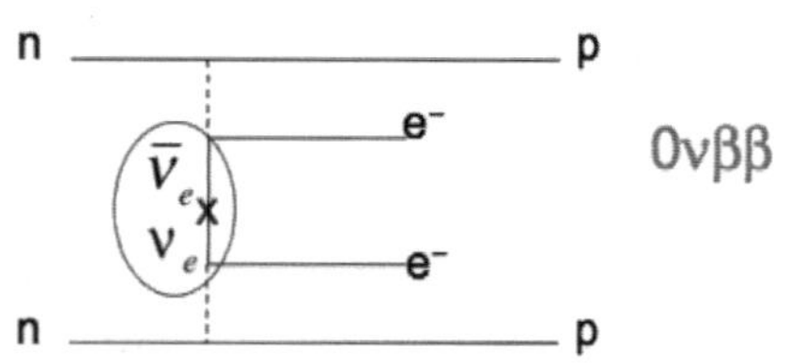

Figure 1. Diagram for neutrinoless double beta decay. It can be seen that the process requires neutrinos to be Majorana fermions. The helicity flip is also depicted.

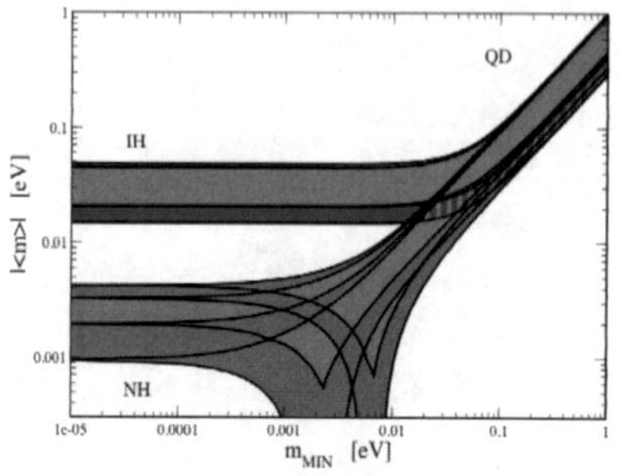

Figure 2. Possible values for the effective Majorana neutrino mass implied by neutrino oscillation data, from [1].

Fig. 2, the best-fit neutrino oscillation parameters (mixing angles and mass-squared differences) were used with ranges $\pm 2\sigma$ for θ_{13}, δ=0, $\pm 1\sigma$ for all other relevant oscillation parameters, and all Majorana phases α from $[0, \pi]$. Blue and green allowed regions (for the inverted and the normal neutrino mass hierarchies, respectively) have CP-conserving Majorana phases, whereas the red allowed regions include at least one of $\alpha_{21}, \alpha_{31}, (\alpha_{31}-\alpha_{21})$ CP-violating.

The full expression for the decay rate for $0\nu\beta\beta$ is:

$$[T_{1/2}]^{-1} = G_{0\nu} \frac{\langle m_{\beta\beta} \rangle^2}{m_e^2} |M_{0\nu}|^2 \quad (2)$$

where $G_{0\nu}$ is the phase space integral (that can be calculated precisely) and $|M_{0\nu}|$ is the nuclear matrix element (NME) for which substantial uncertainties exist in the calculations [2].

There are 35 naturally-occurring isotopes that can undergo double beta decay. Table 1 lists all of the double beta decaying isotopes with Q-value greater than 2 MeV. In general, a larger Q-value is desirable for phase space, detectability and for background considerations. Except for ^{96}Zr and ^{110}Pd, there has been recent and active R&D directed at using each of these isotopes in a double beta decay experiment. In this list of isotopes are all of the main candidates for current and upcoming experiments.

When considering the question of whether there is a favoured isotope for double beta decay, given Equation 2, the candidate with the largest phase space and NME product would be preferred. Different nuclear models yield different calculated NMEs and there are the large uncertainties in the NME calculations that must be taken into account. On the other hand, it has been noted by [3] that a plot of specific phase space versus the geometric mean of NME calculations (for the isotopes listed in Table 1), shows a possible anti-correlation in those two quantities. While there is no obvious reason why this should be true, there is likewise no obvious reason to discount this empirical observation. If true, it would suggest that there is no theoretically favoured isotope, as far as predicted decay rate is concerned. A double beta decay experiment would select their preferred isotope solely from experimental considerations.

2. General experimental considerations and common features

In the next generation of neutrinoless double beta decay experiments, three experimental aspects can be considered to be requirements. Double beta decay experiments must have a large quantity

Table 1. Double beta decaying isotopes with Q-value greater than 2 MeV, sorted from highest decay energy to lowest. Values are from the NNDC website (accessed in 04/2014).

isotope	Q-value [MeV]	natural abundance [%]
^{48}Ca	4.27	0.187
^{150}Nd	3.37	5.6
^{96}Zr	3.35	2.8
^{100}Mo	3.03	9.8
^{82}Se	3.00	8.7
^{116}Cd	2.81	7.5
^{130}Te	2.53	34.1
^{136}Xe	2.46	8.86
^{124}Sn	2.29	5.8
^{76}Ge	2.04	7.73
^{110}Pd	2.02	11.7

of isotope, good energy resolution and very low background levels in the region of interest. There is a variety of experimental approaches for double beta decay currently being pursued and each have relative strengths and weaknesses with respect to these three considerations.

Here, I choose to highlight one of the requirements: very low backgrounds. All of the next-generation experiments have worked extensively on background reduction and rejection. Backgrounds from natural radioactivity (including the usual suspects such as U and Th) are not the only concern. Cosmogenic backgrounds will become more and more important to control as the sensitivity of double beta decay experiments improves, and other background levels are reduced. Background rejection techniques will also need to be exploited. Examples of such techniques include: pulse-shape analysis in Ge diodes to determine an event's spatial topology; track identification of the two beta electrons as a way to strongly reject backgrounds; and using delayed coincidence event signatures (such as the well-known ^{214}Bi-Po). Ultimately it is the amount of isotope and the background level in the energy region of interest that directly determine the sensitivity of a neutrinoless double beta decay experiment.

Section 3 will survey the features, status and results from currently running experiments. I will also report on the future plans of these experiments. Section 4 will summarize the features and status of experiments currently under construction.

3. Currently running experiments

3.1. CANDLES

CANDLES III [4] is taking data in the Kamioka Lab. The experiment uses 96 CaF_2 scintillating crystals (305 kg, with 0.187% ^{48}Ca) immersed in liquid scintillator that serves as a veto. The transparent tank, crystals and photomultiplier tubes (PMTs) are in a water tank for shielding. With a Q-value of 4.27 MeV for ^{48}Ca, the highest of all isotopes, CANDLES is less prone to natural radioactivity backgrounds (that are at lower energies). The expected neutrino mass sensitivity of CANDLES III is 0.5 eV. For the next incarnation of the detector, the enrichment of ^{48}Ca is needed and is being developed.

3.2. KamLAND-Zen

KamLAND is a liquid scintillator neutrino detector. By deploying a mini-balloon in the detector, they are able to put xenon-loaded scintillator inside the balloon in order to conduct a neutrinoless double beta decay search with ^{136}Xe. 320 kg of 90% enriched ^{136}Xe is in the experiment and a total amount of 615 kg is in hand [5]. The main featurs of KamLAND-Zen, as the double beta search experiment is called, is that a massive quantity of isotope can be deployed in a low-background environment. The xenon-loaded scintillator can also be purified and the mini-balloon replaced with relatively low cost.

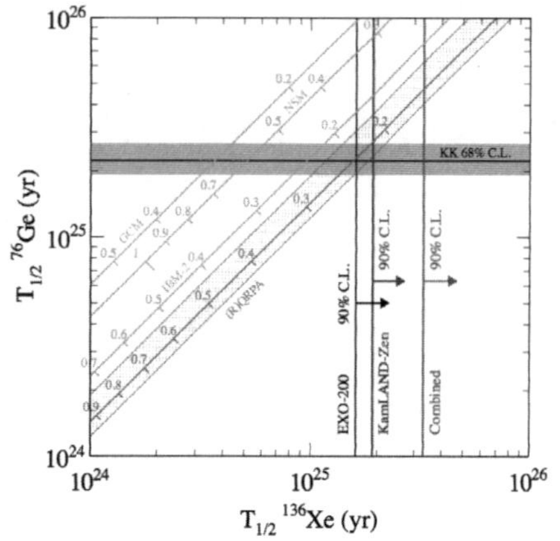

Figure 3. From [6] half-life limits from KamLAND-Zen and EXO-200 compared to the measured half-life claim in [7].

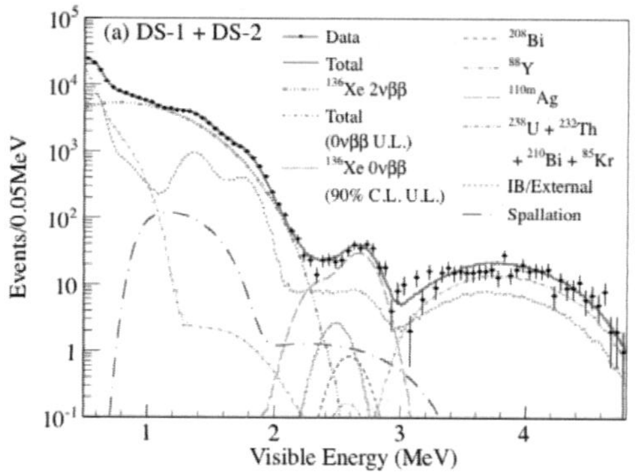

Figure 4. KamLAND-Zen energy spectrum around $Q_{\beta\beta}$ from [6]. The dominant background inferring with the $0\nu\beta\beta$ measurement is likely to be ^{110m}Ag.

Results from the first phase of KamLAND-Zen [6] include a lower limit on the $0\nu\beta\beta$ half-life of $T_{1/2} > 1.9 \times 10^{25}$ yr, corresponding to an upper limit of $\langle m_{\beta\beta} \rangle <$ 120-250 meV, (including the NME model uncertainty). Fig. 3 from [6] shows the half-life limits of KamLAND-Zen and EXO-200, compared to the measured half-life in ^{76}Ge claimed in [7]. The diagonal lines in Fig. 3 correspond to different nuclear theory methods for calculating the NMEs needed to convert half-life limits to effective Majorana neutrino mass; the values shown along the lines are the neutrino masses (in eV). Fig. 4 shows the spectrum. Backgrounds at $Q_{\beta\beta}$ where the $0\nu\beta\beta$ peak is expected include ^{110m}Ag. Observations of the rate of this background over time and comparisons to the half-life of ^{110m}Ag support this conclusion. It is possible that this background was cosmogenic, introduced during the delivery of the xenon.

KamLAND-Zen has completed purifying the xenon and the liquid scintillator in 2013. Data taking has resumed with lower ^{110m}Ag (also lower from decay). A 600 kg deployment with a clean mini-balloon is being planned. There is R&D towards an improved detector with high QE PMTs, higher light yield liquid scintillator and adding light concentrators. Other future possibilities include operating the detector with pressurized xenon (for a higher loading concentration) and R&D toward a scintillating film for the mini-balloon (for background rejection).

3.3. EXO-200

EXO-200 is a liquid xenon TPC that contains 175 kg of 80.6% enriched xenon [8]. It is in the WIPP facility in New Mexico, USA. The key feature of the experiment is that the TPC allows the event topology to be used to very effectively reject multi-site background events (from gamma rays scattering in the detector). Liquid xenon also has effective self-shielding and the combined scintillation and ionization signal offers good energy resolution.

Results from EXO-200 include the first observation of the $2\nu\beta\beta$ decay of ^{136}Xe [9] and a lower limit on the $0\nu\beta\beta$ half-life of $T_{1/2} > 1.6 \times 10^{25}$ yr [8], very similar to that from KamLAND-Zen.

This comes from a 26.3 kg-yr data set taken between 09/2011 to 04/2012. Background control after single-site event selection was very successful with an expectation of 4.1 ± 0.3 background events in a $\pm 1\sigma$ region around $Q_{\beta\beta}$ and an observation of 1 count therein. Their multi-site and single-site energy spectra from [8] are shown in Fig. 5.

The future plans for the experiment are called nEXO and the goal is a 5 tonne LXe TPC as similar to EXO-200 as possible. The desired site for nEXO is the Cryopit in SNOLAB [10]. A five-year run with nEXO aims to reach neutrino mass sensitivity toward the bottom of the inverted hierarchy band (as in Fig. 2).

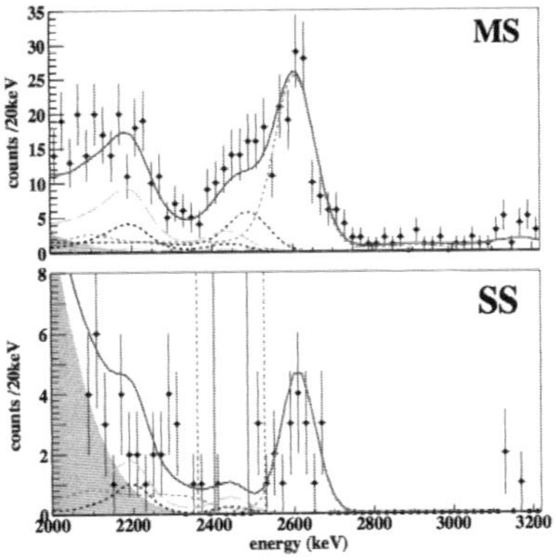

Figure 5. EXO-200 energy spectra around $Q_{\beta\beta}$ from [8]. The substantial reduction of backgrounds achieved by selecting single-site events is evident.

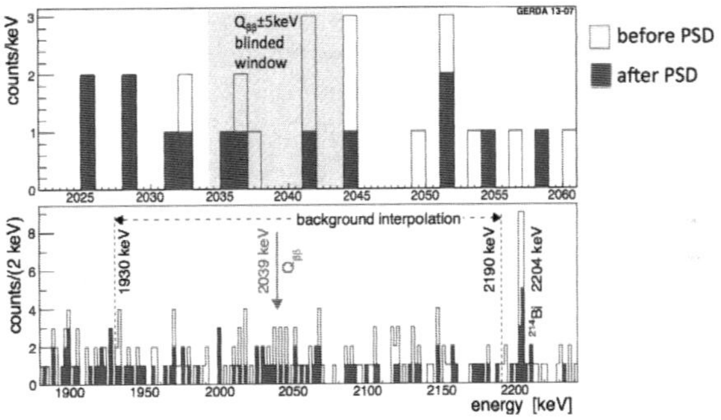

Figure 6. GERDA energy spectrum around $Q_{\beta\beta}$ from [11]. There were 7 events in the blinded window with 3 surviving the PSD cut.

3.4. GERDA

The GERDA experiment [11] in Gran Sasso Lab has operated 18 kg of bare Ge diodes in liquid argon. The Ge detectors are from the Heidelberg-Moscow and IGEX experiments. These detectors are isotopically enriched in ^{76}Ge to 86%. Ge detectors have excellent energy resolution; deployment in liquid argon aims to reduce backgrounds from adjacent detector components. During operation from 11/2011 to 05/2013, two diodes had high leakage current and were shut off; thus, the total mass of enriched detector that was used in the analysis was 14.6 kg.

Pulse-shape analysis was used to reject surface events and also multi-site events from gammas. The data set from a 21.6 kg-yr exposure [12] is shown in Fig. 6. There were 3 counts in the $\pm 2\sigma$ window around $Q_{\beta\beta}$. The expectation for the claimed [7] signal would have been 5.9 ± 1.4 signal events over 2.0 ± 0.3 background events in the same window. The GERDA results strongly disfavour this claim. The lower limit on the half-life of ^{76}Ge is $T_{1/2} > 2.1 \times 10^{25}$ yr at the 90% CL. The GERDA collaboration compares their background levels in a $\pm 2\sigma$ window, noting their value of 0.01 counts/mol-yr versus 0.07 for EXO-200 and 0.2 for KamLAND-Zen.

Phase I of GERDA ended in 09/2013 and all detectors have been dismounted. Phase II involves adding 30 enriched BEGe detectors (a mass of 20 kg). They aim for further background suppression by a factor of 10. This will be achieved by improved radiopurity in the HV and signal cabling, lower radon emanation, improved PSD with the BEGe detectors and instrumenting the liquid argon to function as a scintillation veto.

Journal of Physics: Conference Series **598** (2015) 012008 doi:10.1088/1742-6596/598/1/012008

4. Neutrinoless double beta decay experiments under construction

4.1. CUORE

The CUORE experiment [13] in Gran Sasso Lab is under construction with a test-scale detector CUORE-0 already taking data. CUORE is a cryogenic bolometer detector that will have over 700 kg of TeO_2 crystals containing 206 kg of ^{130}Te isotope. The features of this experiment are the large natural abundance of ^{130}Te and the exquisite energy resolution of a bolometer. There has been a careful approach to background control in CUORE, as is required for an experiment with high sensitivity.

Detector assembly started in 02/2013 and will finish in 2014. Cryostat commissioning and tests are ongoing and full data taking is expected to begin in 2015. Meanwhile CUORE-0 has been taking data since 03/2013. A single CUORE-like tower of 52 TeO_2 bolometers has been installed in the CUORICINO cryostat. This contains 11 kg of ^{130}Te, roughly the same detector mass as in CUORICINO; however the backgrounds in the crystal are expected to be better due to the superior background control efforts used for the CUORE crystals. In particular, the surface alpha backgrounds seen in CUORICINO are expected to be reduced in CUORE, and has been studied with CUORE-0 data.

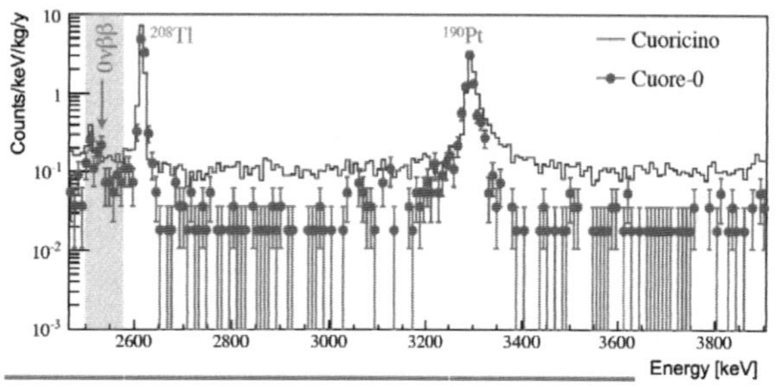

Figure 7. CUORE-0 background levels from [14]. The backgrounds a significantly lower than in CUORICINO.

Figure 8. Photo of copper being electroformed for the MJD detectors from [16].

Fig. 7 from [14] shows the background comparison between CUORE-0 data and CUORICINO. While external gamma backgrounds are expected to be the same (same cryostat), the continuum alpha background in the CUORE crystals is lower due to improved detector surface treatment. The goal for the full CUORE detector, after 5 years of livetime, is an effective Majorana neutrino mass sensitivity at the 47-100 meV level, just touching the top of the inverted mass hierarchy region (shown in Fig. 2).

4.2. MAJORANA

The MAJORANA experiment [15] is aiming for a tonne-scale experiment with enriched Ge detectors. Currently under construction in the Sanford Underground Research Facility in South Dakota, USA is the MAJORANA Demonstrator (MJD). With 30 kg of enriched ^{76}Ge detectors (86% enriched) and 10 kg of ^{nat}Ge detectors, the goal is to demonstrate backgrounds are low enough to proceed with a tonne-scale experiment. The approach is different from GERDA in that a modular array of Ge detectors in ultra-clean copper cryostats is used rather than immersion of bare Ge crystals in liquid argon.

A notable accomplishment of the MJD effort is electroforming of copper underground in Sanford Lab. Fig. 8 shows a photo from [16] of copper being electroformed on a stainless steel mandrel. MJD aims to be taking data in 2014 and will have a second cryostat operational in

2015. The sensitivity of MJD, for a 30 kg-yr exposure and zero background should be in the 100-200 meV range for $\langle m_{\beta\beta} \rangle$.

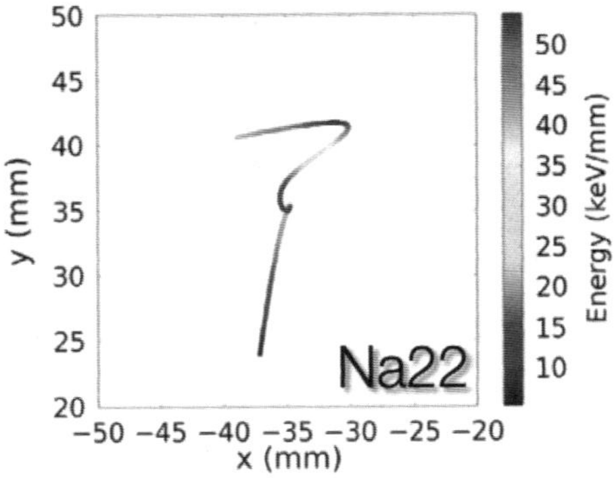

Figure 9. NEXT-DEMO prototype TPC track reconstruction [17].

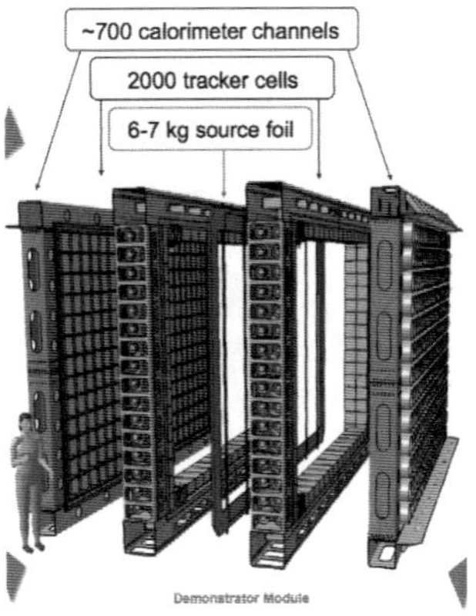

Figure 10. Schematic of a SuperNEMO Demonstrator Module [19].

4.3. NEXT-100

The NEXT collaboration aims to build a 100-150 kg high pressure Xe TPC to search for neutrinoless double beta decay [17]. The plan is to install the detector in the Canfranc underground laboratory in Spain. Enriched xenon gas (90% ^{136}Xe) will be operated at 10-20 bar. Electroluminescence is used to amplify the signals from electrons that have been drifted in order to achieve very good energy resolution. As a gaseous TPC (albeit at high pressure) track reconstruction is an objective for background reduction; the identification of two electron-like tracks in an event will strongly suppress gamma-ray (and other) backgrounds.

The signals in a NEXT TPC would include prompt scintillation (S1) and the electroluminescence (S2) from ionization (drifted electrons). This approach is similar to that used in the two-phase xenon TPC dark matter experiments. In small-scale prototypes, the energy resolution was measured using a ^{22}Na source, and then projected to be 0.7% FWHM at $Q_{\beta\beta}$. Track reconstruction has been demonstrated in prototype detectors, illustrating the charge "blob" at the end of an electron track, as shown in Fig. 9.

Currently under construction is a 10 kg (15 bar) version of the detector called NEXT-White (NEW). The pressure vessel for the TPC has been built and the remaining detector HV and light detection (PMTs and SiPMs) components are being designed and built.

4.4. SuperNEMO

The SuperNEMO experiment is based on the successful NEMO-3 experiment [18]. The detector is a tracking calorimeter which has superb background rejection (using event spatial topology). Source foils of double beta decaying isotopes are deployed in a tracking detector with plastic scintillator calorimeter modules as "book ends" to measure the energy of the electron tracks emitted by the source foil. A schematic of a SuperNEMO detector module is in Fig. 10 [19]. A Demonstrator Module for the full SuperNEMO is under construction and will be deployed in the Modane underground lab in France; it is one module of the full 20 that would constitute SuperNEMO. Each of the 20 modules would have roughly 5 kg of ^{82}Se on the source foil.

The tracker for the Demonstrator Module, being built in the UK, is a 2000-channel Geiger-mode detector. Strict background control is required in its construction and assembly. The energy resolution of the calorimeter readout is expected to be 4% FWHM at $Q_{\beta\beta}$ [19].

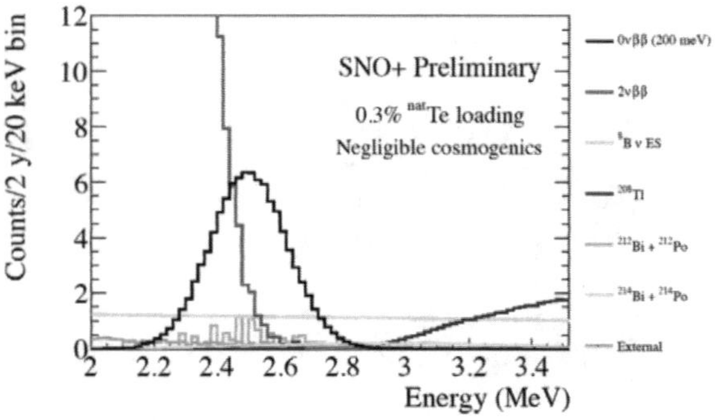

Figure 11. Simulated energy spectrum for SNO+.

Figure 12. Photo of the SNO+ detector filling with water.

4.5. SNO+

The Sudbury Neutrino Observatory (SNO) concluded operations in 2006 and the detector is being refurbished as the SNO+ experiment [20]. With 780 tonnes of liquid scintillator in the detector, SNO+ aims to deploy 0.3% Te, by weight, in the liquid scintillator to conduct a search for neutrinoless double beta decay; this would be 800 kg of ^{130}Te isotope. The main feature of SNO+ is the same as that of KamLAND-Zen; a massive quantity of isotope can be deployed in a low background environment. The SNO+ concept with tellurium goes beyond the 0.3% experiment. I note here that this tonne-scale deployment of Te costs of the order of $1M for the double beta isotope material, as no enrichment is required. This is between 20-100 times cheaper than the cost of a tonne-scale isotope deployment in any other experimental approach. The potential thus exists for the SNO+ approach to make a further step in neutrinoless double beta decay sensitivity, to the 10 tonne scale! Recently, the SNO+ collaboration has succeeded to make 3% Te-loaded liquid scintillator with good light yield.

A liquid scintillator has fast timing that enables good position reconstruction (to suppress external gamma backgrounds) as well as enabling strong background rejection using delayed coincidence events in various decay chains. Factors of several thousand rejection are possible (and have been achieved in KamLAND-Zen). With a liquid scintillator detector, energy resolution is not as good as in other detector technologies. It is an important fact that the $2\nu\beta\beta$ rate in ^{130}Te is one of the slowest of all of the candidate isotopes for a double beta decay experiment; this helps minimize the background from $2\nu\beta\beta$ that tails into the energy region at $Q_{\beta\beta}$. With this fact, plus purification for reducing backgrounds and background rejection techniques, SNO+ has the goal that in the energy region of interest for $0\nu\beta\beta$, corresponding roughly to $[-0.5\sigma, 1.5\sigma]$ around $Q_{\beta\beta}$, the dominant background will be ^{8}B solar neutrinos.

SNO+ has studied the purification of tellurium and Te-loaded liquid scintillator, including a thorough study of possible cosmogenic backgrounds. The expected backgrounds and signals in SNO+ are shown in Fig. 11. With 0.3% Te loading, SNO+ sensitivity to $\langle m_{\beta\beta}\rangle$ is expected to reach as low as 50 meV. The successful demonstration of background control at this level would pave the way for a 3% Te SNO+ experiment with $\langle m_{\beta\beta}\rangle$ sensitivity reaching to the bottom of the inverted mass hierarchy allowed region. SNO+ is currently under construction. The detector is being filled with water (see Fig. 12) and the scintillator purification plant is being installed. Data taking with scintillator is expected to begin in 2015.

Acknowledgments
I wish to thank all the representatives of the experiments who provided me with material to present at NuPhys 2013. Citations have been included in this report. In addition, communications and information were received from R. Saakyan, F. Avignone, E. Fiorini, O. Cremonesi, K. Kumar, J. Farine and S. Schönert.

References
[1] Pascoli S updated figure in RPP 2013 "Neutrino Mass, Mixing and Oscillations", originally in Pascoli S and Petcov S 2008 *Phys. Rev.* D **77** 113003
[2] Šimkovic F 2014 *J. Phys.: Conf. Series* this volume
[3] Robertson R G H 2013 *Mod. Phys. Lett.* a **28** 1350021
[4] Umehara S and Kishimoto T 2013, private communication
[5] Inoue K 2013, private communication
[6] Gando A *et al.* 2013 *Phys. Rev. Lett.* **110** 062502
[7] Klapdor-Kleingrothaus H V and Krivosheina I V 2006 *Mod. Phys. Lett.* A **21** 1547
[8] Auger M *et al.* 2012 *Phys. Rev. Lett.* **109** 032505
[9] Ackerman N *et al.* 2011 *Phys. Rev. Lett.* **107** 212501
[10] Gratta G 2013, private communication
[11] Ackermann K-H *et al.* 2013 *Eur. Phys. J.* C **73** 2330
[12] Agostini M *et al.* 2013 *Phys. Rev. Lett.* **111** 122503
[13] Artusa D R *et al.* 2014 Searching for neutrinoless double-beta decay of ^{130}Te with CUORE *Preprint* arXiv:1402.6072
[14] Brofferio C 2013 private communication, subsequently in Aguirre C P *et al.* 2014 Initial performance of the CUORE-0 experiment *Preprint* arXiv:1402.0922
[15] Martin R D *et al.* 2013 Status of the MAJORANA DEMONSTRATOR experiment *Preprint* arXiv:1311.3310
[16] Elliott S R 2013, private communication
[17] Gómez-Cadenas J J 2013, private communication
[18] Arnold R *et al.* 2005 *Nucl. Instrum. Meth.* A **536** 79
[19] Waters D 2013, private communication
[20] Chen M C 2008 The SNO+ experiment, proceedings of the 34th International Conference on High Energy Physics (ICHEP 2008) *Preprint* arXiv:0810.3694

Neutrino Masses at the LHC

Frank F. Deppisch

Department of Physics and Astronomy, University College London, UK

E-mail: f.deppisch@ucl.ac.uk

Abstract. Neutrinoless double beta decay is the traditional tool to probe Majorana neutrino masses and lepton number violating physics in general. On the other hand, many models incorporating Majorana neutrino masses also predict new states and lepton number violating interactions at the TeV scale that can potentially be probed at the LHC. We provide a brief overview of the pertinent operators and a selection of physics models in order to highlight the interplay between neutrinoless double beta decay and LHC searches.

1. Introduction

Neutrinoless double beta decay ($0\nu\beta\beta$) is the most sensitive tool for probing Majorana neutrino masses. However, while the so-called mass mechanism is certainly the best known example triggering the decay, Majorana neutrino masses are not the only element of beyond Standard Model physics which can induce it. In this proceedings report we highlight possible other mechanisms of $0\nu\beta\beta$ decay where the lepton number violation (LNV) does not directly originate from the exchange of light Majorana neutrinos but rather due to LNV in extensions of the Standard Model. Many of these extensions predict new states and lepton number violating interactions at the TeV scale that can be potentially probed at the LHC. Correlating $0\nu\beta\beta$ with searches at the LHC provides a powerful tool to distinguish between different LNV sources and mechanisms of neutrino mass generation.

In this report, we provide a brief overview of the possible effective operators (c.f. Figure 1) that can trigger $0\nu\beta\beta$ beta decay, and discuss example signatures of lepton number and flavour violation at the LHC. For more details, see the review [1] and references therein.

2. Neutrinoless Double Beta Decay

2.1. Standard Mass Mechanism

Recall that the standard light neutrino exchange of $0\nu\beta\beta$ probes the effective mass $\langle m_\nu \rangle = \sum_j U_{ej}^2 m_j \equiv m_{ee}$, where the sum is over the active light neutrinos. The inverse $0\nu\beta\beta$ half life in a given isotope is then $[T_{1/2}^{0\nu\beta\beta}]^{-1} = |\langle m_\nu \rangle / m_e|^2 G_0 |ME|^2$, where G_0 and $|ME|$ denote the nuclear phase space factor and matrix element, respectively. The non-observation of $0\nu\beta\beta$ in current experiments points to a limit $\langle m_\nu \rangle \lesssim 0.5 - 1.0$ eV.

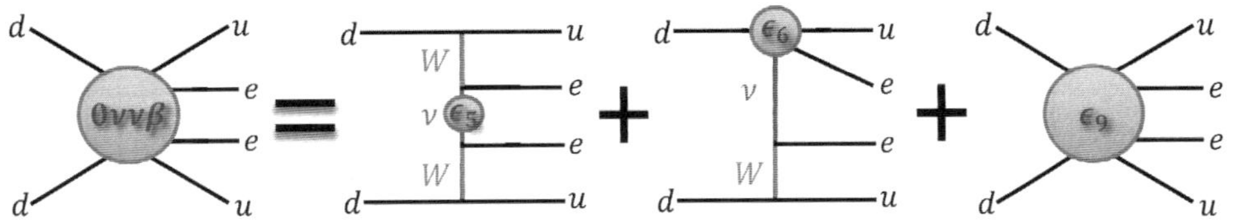

Figure 1. Schematic overview of different contributions to $0\nu\beta\beta$: Standard mass mechanism, long–range 6-dim operator, short–range 9-dim operator.

2.2. Long–Range Contributions

Long–range contributions to $0\nu\beta\beta$ decay involve two effective vertices connected by the exchange of a light neutrino. The general Lagrangian is expressed in terms of effective couplings ϵ_β^α [2],

$$\mathcal{L} = \frac{G_F}{\sqrt{2}} \left(j^\mu_{V-A} J^\dagger_{V-A,\mu} + \sum_{\alpha,\beta} \epsilon^\beta_\alpha j_\beta J^\dagger_\alpha \right), \tag{1}$$

with the hadronic and leptonic currents $J^\dagger_\alpha = \bar{u}\mathcal{O}_\alpha d$ and $j_\beta = \bar{e}\mathcal{O}_\beta \nu$, respectively. The sum is over all Lorentz-invariant combinations, except for the standard case $\alpha = \beta = (V - A)$, and all currents are scaled relative to the Fermi interaction strength. The individual operators $\mathcal{O}_\alpha$ are

$$\mathcal{O}_{V\pm A} = \gamma^\mu(1 \pm \gamma_5), \qquad \mathcal{O}_{S\pm P} = (1 \pm \gamma_5), \qquad \mathcal{O}_{T_\pm} = \frac{i}{2}[\gamma_\mu, \gamma_\nu](1 \pm \gamma_5). \tag{2}$$

The interpretation of the effective couplings ϵ_β^α depends on the specific particle physics model. Assuming the dominance of one of the couplings, the inverse $0\nu\beta\beta$ half life is

$$[T^{0\nu\beta\beta}_{1/2}]^{-1} = |\epsilon^\beta_\alpha|^2 G_{0k} |ME|^2, \tag{3}$$

where G_{0k} and $|ME|$ denote the corresponding nuclear phase space factor and matrix element, respectively. Current limits on the effective couplings are of the order $\epsilon \lesssim 10^{-9} - 10^{-7}$, depending on the operator [1]. This points to a limit on the scale of LNV physics of $\Lambda_{\rm LNV} \gtrsim 10^6$ GeV, generating the 6-dimensional operators in (1).

2.3. Short–Range Contributions

Short–range contributions to $0\nu\beta\beta$ decay involve one effect vertex originating from the general Lorentz-invariant Lagrangian [3]

$$\mathcal{L} \;=\; \frac{G_F^2}{2m_p}\left(\epsilon_1 JJj + \epsilon_2 J^{\mu\nu}J_{\mu\nu}j + \epsilon_3 J^\mu J_\mu j + \epsilon_4 J^\mu J_{\mu\nu} j^\nu + \epsilon_5 J^\mu J j_\mu\right), \tag{4}$$

consisting of 9-dimensional effective operators, with the hadronic currents $J = \overline{u}(1 \pm \gamma_5)d$, $J^\mu = \overline{u}\gamma^\mu(1 \pm \gamma_5)d$, $J^{\mu\nu} = \overline{u}\frac{i}{2}[\gamma^\mu, \gamma^\nu](1 \pm \gamma_5)d$ and the leptonic currents $j = \overline{e}(1 \pm \gamma_5)e^C$, $j^\mu = \overline{e}\gamma^\mu(1 \pm \gamma_5)e^C$. The $0\nu\beta\beta$ decay rate can be expressed as in (3) with the corresponding phase space factors and matrix elements. Similar to the long-range operators case, current limits on the effective couplings are of the order $\epsilon \lesssim 10^{-9} - 10^{-7}$, depending on the operator [1]. In this case, the scale of LNV physics probed is of the order $\Lambda_{\rm LNV} \gtrsim 1$ TeV. This hints at the possibility that physics generating these operators can be probed at the LHC as well.

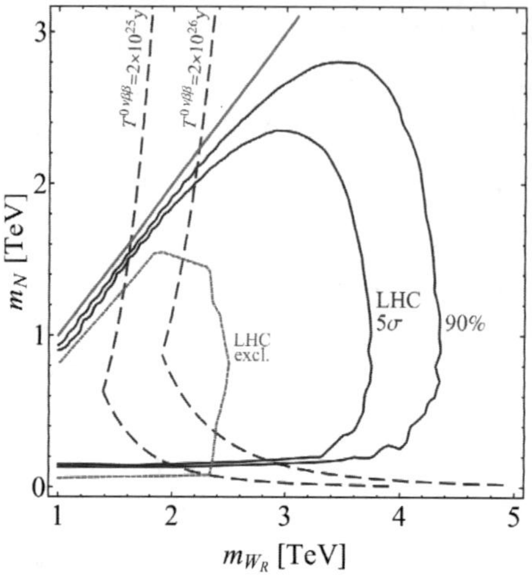

Figure 2. Comparison of LNV event rates at the LHC and in $0\nu\beta\beta$ experiments [9]. The solid blue contours give the signal significances of 5σ and 90% at the LHC with 14 TeV and $\mathcal{L} = 30$ fb^{-1}. The red shaded area is excluded by current LHC searches [10]. The green dashed contours show the sensitivity of current and future $0\nu\beta\beta$ experiments, assuming dominant doubly-charged Higgs or heavy neutrino exchange.

3. Lepton Number Violation at the LHC

3.1. Left-Right Symmetry

As an example of a model incorporating the rich phenomenology of lepton number violation, we briefly discuss the minimal Left-Right symmetric model (LRSM) which extends the Standard Model gauge symmetry to the group $SU(2)_L \otimes SU(2)_R \otimes U(1)_{B-L}$ [4, 5, 6, 7, 8]. Lepton number violation and the presence of right-handed neutrinos are a necessary ingredient. The LRSM accommodates a general Seesaw type I + II neutrino mass matrix.

3.2. Neutrinoless Double Beta Decay

The model provides several mechanisms that contribute to $0\nu\beta\beta$ decay: (i) Standard light neutrino exchange with mass helicity flip; (ii) Long-range light neutrino exchange with right-handed currents; (iii) Short-range heavy right-handed neutrino exchange; (iv) Short-range right-handed doubly-charged triplet Higgs exchange. Each of these contributions can be mapped to one of the effective operators discussed in the previous section. For example, $0\nu\beta\beta$ decay through the exchange of heavy neutrinos with right-handed currents via the right-handed W_R gauge boson corresponds to the effective coupling

$$\epsilon_3^{RR} = \sum_{i=1}^{3} V_{ei}^2 \frac{m_p}{m_{N_i}} \frac{m_{W_L}^4}{m_{W_R}^4}, \tag{5}$$

with the heavy neutrino masses N_i and the mixing matrix between the heavy neutrinos V_{ei}.

3.3. Lepton Number and Flavour Violation at the LHC

In the LRSM, lepton number violation can also be probed via heavy right-handed neutrino exchange leading to the signal $pp \to W_R \to e^{\pm}\mu^{\pm,\mp} + 2$ jets at the LHC [11]. The potential to discover lepton flavour and lepton number violation using this process has been analyzed

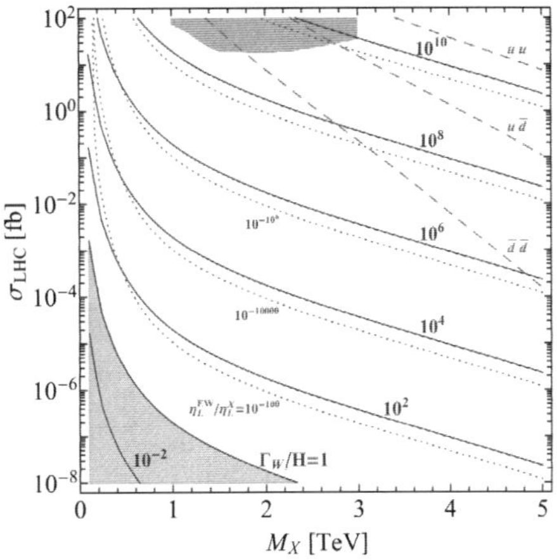

Figure 3. Lepton number washout rate Γ_W/H at $T = M_X$ as a function of the LNV scale M_X and the corresponding LHC cross section σ_{LHC} (solid blue contours). The dotted light blue contours denote the lepton asymmetry at the electroweak scale relative to its value at M_X, $\eta_L^{\mathrm{EW}}/\eta_L^X$. The red dashed curves are typical cross sections of the process $pp \to l^{\pm}l^{\pm}qq$. The shaded region at the top is excluded by recent searches at the LHC.

in [12, 9]. Figure 2 compares the LNV event rate at the LHC with the sensitivity of $0\nu\beta\beta$ experiments. The green dashed contours represent the excluded areas from $0\nu\beta\beta$ searches using nominal values for the current and future sensitivity. In this analysis it was assumed that $0\nu\beta\beta$ is dominated by either heavy neutrino or Higgs triplet exchange. As the contribution from the standard light neutrino exchange is always present, this corresponds to a scenario with a small effective mass $\langle m_\nu \rangle$. Figure 2 provides an example of the potential synergy between LNV searches at the LHC and in $0\nu\beta\beta$ experiments.

3.4. Falsifying Leptogenesis

The observation of lepton number violation at the LHC would not only have important consequences on the physics of $0\nu\beta\beta$ and neutrino mass generation but also on the viability of Leptogenesis models. In the traditional Leptogenesis scenario, the observed baryon asymmetry of the Universe is explained through the generation of a net lepton number asymmetry through the out-of-equilibrium decay of heavy right-handed neutrinos, which is then converted to the matter asymmetry though Standard Model $(B + L)$-violating sphaleron interactions. As part of this mechanism, the generation of a net lepton number asymmetry has to be balanced with necessarily present processes that washout this asymmetry.

Observing any lepton number violating process at the LHC would put a strong lower limit on this washout rate. This relation between general LNV processes at the LHC and the effect on leptogenesis models was discussed in [13]. The minimal rate of washout as a function of the LHC cross section and the scale of lepton number violation (for example observed as a resonance at this mass) is shown in Figure 3. For a washout rate larger than one, the dilution of lepton number is highly effective and a model that generates a lepton number asymmetry above the TeV scale is not a viable scenario to explain the observed baryon asymmetry. As shown in the figure, any observation of LNV at the LHC would correspond to a much larger washout rate, and would therefore strongly constrain Leptogenesis scenarios above the scale M_X. Low scale scenarios, such as resonant Leptogenesis where the lepton asymmetry is generated at scales lower

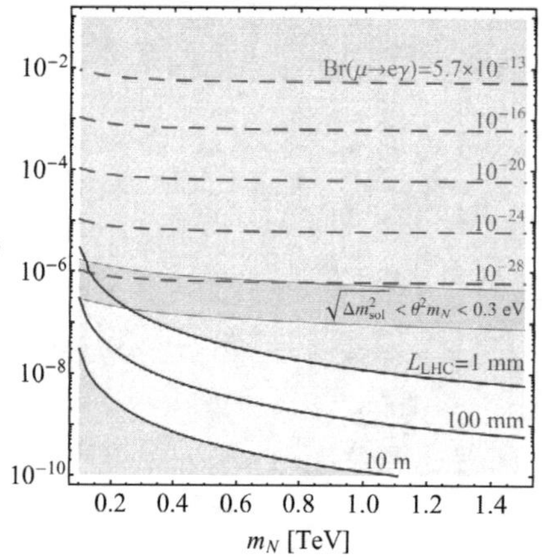

Figure 4. Average decay length of a heavy neutrino as a function of its mass m_N and the light-heavy mixing θ (solid blue contours). The dashed red contours denote $Br(\mu \to e\gamma)$ whereas the grey shaded band corresponds to light neutrino mass scales $m_\nu = \theta^2 m_N$ between $\sqrt{\Delta m^2_{\text{sol}}}$ and 0.3 eV within the canonical type-I seesaw mechanism (taken from [14]).

than M_X would not necessarily be constrained.

3.5. Probing Small Yukawa Couplings

While total lepton number violation is considered a smoking gun signal of the Majorana nature of neutrinos, the observed neutrino oscillations already provide clear evidence of individual lepton flavour violation (LFV). Searches for rare processes such as $\mu \to e\gamma$ already put stringent limits on charged lepton flavour violation, which usually prohibits the observation of flavour violation at the LHC. This can be evaded in a scenario where right-handed neutrinos are produced via a Z' portal but which only decay via small flavour violating couplings θ [14]. The LHC process rate is then not suppressed by the small LFV despite unobservably small $\mu \to e\gamma$. The process under consideration is $pp \to Z' \to NN \to e^\pm\mu^\mp + 4j$, through the resonant production of two heavy neutrinos via a Z' portal. The neutrino N can decay via the channels $\ell^\pm W^\mp$, $\nu_\ell Z$ and $\nu_\ell h$, all of which are suppressed by the small light-heavy neutrino mixing θ. As long as the total neutrino decay width is large enough for decays within the detector, the process rate is not suppressed by the overall mixing strength θ. The decay length of the heavy neutrino is shown in Figure 4 (left), in relation with $Br(\mu \to e\gamma)$. It demonstrates that the canonical Seesaw type I regime with TeV scale neutrinos and small Yukawa couplings $\theta \approx 10^{-6}$ cannot be probed by low energy searches but potentially by the LHC process discussed here.

4. Summary

The Standard Model of particle physics (here defined as including light neutrino masses) has so far evaded all attempts to unambiguously prove it wrong. The discovery of a SM compatible Higgs boson at the LHC provides the most recent and dramatic confirmation of its predictions. Neutrinos provide a natural channel to look for (further) signs beyond the Standard Model as they are still the least understood matter particles; even if the Higgs mechanism of fermion mass generation is confirmed, we do not understand why neutrinos are as light as they are. Of crucial importance in this regard is not only the value of the mass of neutrinos but also their nature: Dirac or Majorana? Neutrinoless double beta decay is the most important observable to probe for both. In addition, it tests the fundamental symmetry of lepton number. Despite the lack of any signs of new physics, the LHC does and most importantly will provide important information in this regard as it probes models of neutrino mass generation and lepton number violation at the TeV scale. In this report, we briefly highlighted the connection between neutrinoless double beta decay and searches for lepton number and flavour violating processes at the LHC. The observation of lepton number violation at the LHC would have a dramatic impact on aspects of neutrino physics such as Leptogenesis, but even if nothing was to be observed, it would affect our understanding of neutrinoless double beta decay.

References

[1] Deppisch F F, Hirsch M and Pas H 2012 *J.Phys.* **G39** 124007 (*Preprint* 1208.0727)
[2] Päs H, Hirsch M, Klapdor-Kleingrothaus H and Kovalenko S 1999 *Phys.Lett.* **B453** 194–198
[3] Päs H, Hirsch M, Klapdor-Kleingrothaus H and Kovalenko S 2001 *Phys.Lett.* **B498** 35–39 (*Preprint* hep-ph/0008182)
[4] Pati J C and Salam A 1974 *Phys.Rev.* **D10** 275–289
[5] Mohapatra R and Pati J C 1975 *Phys.Rev.* **D11** 2558
[6] Senjanovic G and Mohapatra R N 1975 *Phys.Rev.* **D12** 1502
[7] Mohapatra R N and Senjanovic G 1981 *Phys.Rev.* **D23** 165
[8] Duka P, Gluza J and Zralek M 2000 *Annals Phys.* **280** 336–408 (*Preprint* hep-ph/9910279)
[9] Das S, Deppisch F, Kittel O and Valle J 2012 *Phys.Rev.* **D86** 055006 (*Preprint* 1206.0256)
[10] Aad G *et al.* (ATLAS Collaboration) 2012 9 pages plus author list (23 pages total), 4 figures, 3 tables, submitted to European Physical Journal C (*Preprint* 1203.5420)
[11] Keung W Y and Senjanovic G 1983 *Phys.Rev.Lett.* **50** 1427
[12] Aguilar-Saavedra J, Deppisch F, Kittel O and Valle J 2012 *Phys.Rev.* **D85** 091301 (*Preprint* 1203.5998)
[13] Deppisch F F, Harz J and Hirsch M 2013 (*Preprint* 1312.4447)
[14] Deppisch F F, Desai N and Valle J W F 2014 *Phys.Rev.* **D89** 051302 (*Preprint* 1308.6789)

Neutrino properties from cosmology

Licia Verde

ICREA & Instituto de Ciencias del Cosmos, Universidad de Barcelona ICC (UB-IEEC), Marti i Franques 1, 08028, Barcelona, Spain
&
Institute for Theoretical Astrophysics, University of Oslo, 0315 Oslo, Norway.

E-mail: licia.verde@icc.ub.edu

Abstract. In the past few years there have been new developments in the effort of constraining neutrino properties with cosmology. The Cosmic Microwave Background has been measured with renewed and improved precision and large-scale structure surveys have mapped cosmological structures in the Universe over unprecedentedly large volumes. Future, massive large-scale structure surveys have been presented and approved. On the theory side, a significant effort has been devoted to achieve better modelling of small scale clustering and of cosmological non-linearities. As a result it has become clear that forthcoming cosmological data have, in principle, enough statistical power to detect the effect of non-zero neutrino mass (even at the lower mass scale limit imposed by oscillations) and to constrain the absolute neutrino mass scale. I will present some recent work on constraints on neutrino properties from cosmology, concentrating in particular on the work done by my group and my collaborators.

1. Introduction

Cosmology has witnessed an avalanche of data over the past decade or so which has marked the inception of "precision cosmology". A standard model for cosmology has been confirmed: this model with only few parameters describes well observations of the Universe over a wide range of scales and epochs. This model, however, is a model, which, we know, cannot be a complete description and, we know, it is incomplete. Thus by entering in the era of precision cosmology, the focus has shifted from measuring the parameters of the model, to looking for deviations for it and testing the physics on which it is based. Among the many areas of active research in this direction is the goal of studying neutrino properties with cosmology.

Experimental cosmology is however a completely different endeavour from other branches of experimental physics as we only have one observable Universe. We can only make observations (only of the observable Universe) and not controlled experiments; we then fit models (i.e., constrain cosmological parameters) to observations. Any statement or constraint is therefore model-dependent.

Moreover, while the physics of the early Universe is relatively simple and well understood, complicated non-linear physics and astrophysics drive observable quantities in the late-time Universe. Different observations are subject to different levels of trust and robustness because of the variety of poorly known astrophysical processes involved and other systematic effects. Any statement or constraint therefore depends also on the data set chosen. (A practical example will be discussed in §5).

This is what I call the "curse" of cosmology. It is also somewhat a blessing in the sense that in principle we can observe all there is to observe and therefore make "ultimate" experiments: taking more data will not improve the measurement for an "ultimate" experiment, there is a statistical error-floor. Observational cosmology is now getting to the point that for some observable quantities "ultimate" experiments are being made and the statistical error-floor is being reached.

These considerations are extremely important especially when trying to learn about neutrino properties from cosmological observations. In particular, all signals are indirect signals and must be interpreted in the framework of a cosmological model (with a watchful eye for residual systematic and astrophysical effects).

2. Neutrinos signatures in Cosmology

What is a neutrino for cosmology? It is something that behaves like radiation at $T \sim eV$ corresponding to the recombination and decoupling epoch, eventually (possibly, but possibly not) becomes non-relativistic, and behaves like matter. When this happens, it must have small interactions and therefore does not behave like a perfect fluid, and has a high velocity dispersion –it is "hot"–. Of course, neutrinos with masses below $\mathcal{O}(10)$ eV do fit the bill, but one can imagine other processes that also would.

Cosmologically relevant neutrinos are called relict neutrinos and belong to the Cosmic Neutrino Background, which is a relict of the big bang much like the Cosmic Microwave Background (CMB), but which decouples from matter some 2 s after the big bang and not $380,000$ years. Neutrinos are in equilibrium with the primeval plasma through weak interactions, and decouple from it at a temperature $T \sim 1$MeV. At decoupling neutrinos are still relativistic (if, not they would essentially behave like dark matter for cosmological observations and would not be distinguishable from it) and have large velocity dispersions.[1] The minimum total neutrino mass inferred by neutrino oscillations results indicates that neutrinos contribute at least to 0.5% of the total matter density. In the era of precision cosmology even this half % is non-negligible! As such neutrinos affect the growth of cosmic clustering so they can leave imprints on cosmological observables.

It is particularly interesting to resort to cosmology to constrain neutrino properties because cosmology is sensitive to the absolute neutrino mass scale, and current cosmological constraints on the sum of neutrino masses $\sum m_\nu$ are competitive with forecasted bounds from ambitious terrestrial experiments (see e.g., the Katrin experiment[2]).

We have essentially three windows to study relict neutrinos: at primordial nucleosynthesis $T \sim$ MeV, neutrinos behave like radiation, any deviation from the standard 3 neutrino families would change the Universe expansion rate[3], changing the abundances of primordial light elements. This window probes the number of neutrino species (or effective number of species). Here I will concentrate on the other two windows: the CMB and the Large-scale clustering. These observables can be used to probe both number of species and masses.

3. Constraining or measuring neutrino masses

Neutrinos with total masses $\gtrsim$1 eV become non-relativistic before recombination and therefore their effects can be seen in the (primary) CMB. Neutrinos with $\sum m_\nu \lesssim 1$ eV still alter matter

[1] There are about 100 relict neutrinos per cubic centimetre, to be compared with 60 Billion solar neutrinos per second per square cm. Relict neutrinos are of much lower energy but are distributed throughout the entire Universe.

[2] https://www.katrin.kit.edu/68.php

[3] Recall that while matter density scales with redshift, z, as $(1+z)^3$, radiation and relativistic particles density scale like $(1+z)^4$. The radiation content is extremely well known because the CMB temperature is exquisitely measured.

radiation equality but this effect on the CMB can be "cancelled" through degeneracies with other cosmological parameters. On the other hand, due to their high velocity dispersion neutrinos free stream. Therefore finite neutrino masses suppress the matter power spectrum on scales smaller than the free streaming length. In linear theory (for the growth of cosmological perturbations) the suppression flattens out at scales $k > 0.1h/\text{Mpc}$: it is a 50% effect for $\sum m_\nu = 1$ eV a 15% for $\sum m_\nu = 0.3$ eV and a 6% effect for $\sum m_\nu = 0.3$ eV (this can be seen in the solid line of Fig. **??** (from [3]) showing the fractional change in the matter power spectrum as function of wave-number k. In addition different masses become non-relativistic at slightly different times affecting the shape of the matter power spectrum (more in §6).

The recently announced results from the Planck satellite place constraints at $\sum m_\nu < 0.66$ eV at the 95% confidence from CMB only and at $\sum m_\nu < 0.25$ eV when also considering constraints on the lower redshift expansion history coming from the Baryon Acoustic Oscillation (BAO) measurements which help break cosmological parameters degeneracies [1]. This constraint is interesting as it disfavours the degenerate hierarchy. It also puts cosmological (95% confidence) constraints on the sum of neutrino masses at a value just below the Katrin experiment window for detection–90% limit. Needless to say that it will be very interesting to compare the constraints from the two approaches when available.

To improve on the CMB constraint we must look at the shape of the matter power spectrum and thus at the clustering of large-scale structure. It is important to note that the effect of non-zero neutrino masses on the matter power spectrum is not small. A few % amplitude localised features in the power spectrum (the Baryon Acoustic Oscillations, or BAO) have been measured with high confidence! However the neutrino signature is a broadband feature and thus more prone to systematic errors. Statistical errors on the measurement of the power spectrum scale like the square root of the volume surveyed and planned surveys aim at covering a large fraction of the observable Universe. Forecasts for the Euclid satellite based on linear theory for the growth of cosmological perturbations [2] show that such survey should be able to see the signature of neutrino masses in the sky even for $\sum m_\nu$ at the lower limit imposed by oscillations. The main limitation to reach this goal is systematics: the power spectrum of the dark matter cannot be directly measured, in addition at small scales (where there is a lot of signal-to-noise) poorly known astrophysical and baryonic effects kick in. It would be great to have an internal consistency check in case of a detection of $\sum m_\nu > 0$. I will return to this later. The main limitation of linear theory predictions are the effects of non-linearities: non-linear scales enclose a lot of signal-to-noise (noise is scaling like the maximum wavenumber to the 3/2 power). Over the past three years a lot of advances have been made in modelling non-linear gravitational evolution in the presence of massive neutrinos. Approaches that have been considered include analytic/perturbation theory modelling, N-body simulations which also model neutrinos in a fully non-linear way, and an intermediated approach where neutrinos are evolved linearly and therefore analytically but the dark matter is simulated via N-body. Approaches to simulate both dark matter and neutrino masses include using particles to follow neutrinos, using grids or using an hybrid approach. Our group has simulated also the effects of mass hierarchy. Each of the approaches has of course its own advantages and disadvantages, but, encouragingly, the different approaches are in good agreement. The overall conclusion is that non-linearities enhance the signal compared to linear theory predictions both for the effect of neutrino masses and the effect of the hierarchy (to be discussed in §6). Figure 1 show the effects non-zero neutrino masses on the matter power spectrum at non-linear scales (figure from [3]). Here and hereafter NH denotes the normal mass hierarchy and IH denotes the inverted mass hierarchy.

However these predictions are for dark matter only and there are a lot of real-world effects that might make these scales and these gains not reachable. These effects are baryonic physics (which affects all probes including weak gravitational lensing), bias and redshift space distortions (both of which affect galaxy surveys). This is an active research area and no consensus has been

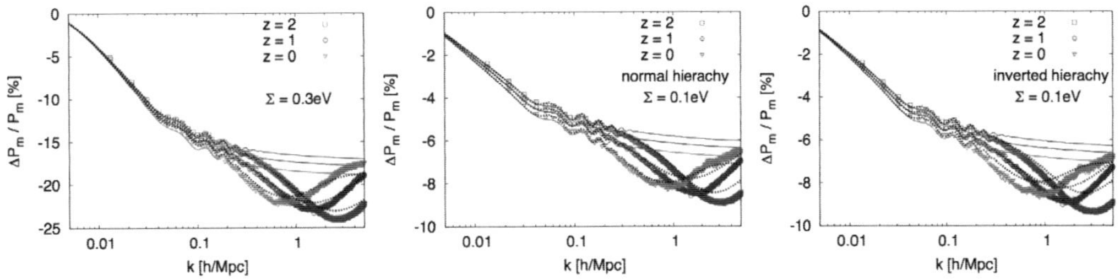

Figure 1. Fractional difference in the total matter power spectrum of the N-body runs with massive neutrinos and the masses neutrino run. Data points show the simulation results, the solid line linear theory predictions and black dotted lines show the estimated effect including non-linearities using the extended HALOFIT formula [4]. Note the typical shape of the effect in the non-linear case: non-linearities enhance the effect and lead to a maximum in the suppression located at mildly non-linear scales.

reached.

On the other hand a lot of these limitations could be bypassed by resorting to the imprint of the growth of clustering on the CMB radiation via gravitational lensing of the CMB. Gravitational lensing by large scale structure of the CMB light, modifies coherently not just the CMB temperature pattern but also the polarisation. This signal was undetected until few years ago, but since then enormous advances have been made. From the first detection in 2007 ([5]) exploiting the cross correlation of the NVSS survey (a large scale structure tracer) and the CMB temperature, to a $> 20\,\sigma$ detection from the Planck temperature maps alone via the the imprint lensing leaves in the temperature four point function, see [6] and refs therein. In the past few months we have seen detection of the signal also in polarisation [7]. CMB lensing is quite sensitive to neutrino masses and given how rapidly this field is evolving we might expect powerful new constraints soon.

Since the publication of the Planck results and constraints in March 2013, many papers have appeared claiming detection of non-standard neutrino properties, $\sum m_\nu > 0$ and/or more than three neutrino species [8, 9, 10, 11, 12]. Before the publication of the Planck results there were few claims for extra neutrinos e.g., [13] and the compilation by [14]. These findings are all somewhat related and will be discussed below, §5.

4. Number of neutrino species

In the standard model there are three neutrino families but in principle there could be more (sterile neutrinos). Free-streaming relativistic particles affect the CMB in two ways *1)* through their relativistic energy density, which alters the epoch of matter-radiation equality, boosting the expansion rate, and *2)* through their anisotropic stress. The first effect however can be mimicked by any other process that alters the expansion rate. For this reason the relevant parameter is called effective number of neutrino species, $N_{\rm eff}$. The fiducial $N_{\rm eff}$ value for three neutrino families is slightly higher (3.046) to account for a small neutrino heating during $e^+ e^-$ annihilation. Assuming we keep fixed that angular size distance to the CMB (i.e., the position of the first peak), the redshift of matter-radiation equality and the physical baryon density the change of expansion rate due to $N_{\rm eff}$ affects both the distance acoustic waves travel $\propto t \propto H^{-1}$ and the distance photons diffuse $\propto t \propto H^{-1/2}$ giving a net overall effect of increasing Silk damping on small CMB scale. An increase in $N_{\rm eff}$ increases the expansion rate of the Universe at all redshifts affecting the Hubble constant H_0 and age of the Universe. Moreover, by affecting matter radiation equality and radiation drag, $N_{\rm eff} \neq 3.046$ affects the baryon acoustic scale and the interpretation of BAO measurements.

While the first effect gives a degeneracy between $\Omega_m h^2$ and $N_{\rm eff}$ and H_0 and $N_{\rm eff}$, the second effect can break this degeneracy somewhat and affects the CMB damping tail. Neutrino perturbations cut effectively the degeneracy at low $N_{\rm eff}$ (some neutrinos are needed to explain the peaks height and location) but the degeneracy extends to high $N_{\rm eff}$.

5. Claims of detection of non-standard neutrino properties.

Before the advent of Planck data, the CMB degeneracy in $N_{\rm eff}$-H_0 plane extended to high values of $N_{\rm eff}$ and H_0. The fact that local H_0 measurements cut this degeneracy at $N_{\rm eff} \gtrsim 3$ was driving all pre-Planck claims of sterile neutrinos from cosmology. This is discussed in details in [15].

With Planck data and a much improved measurement of the CMB damping tail the CMB-only constraint became $N_{\rm eff} = 3.36^{+0.68}_{-0.64}$ (95% CL), including BAO measurements which break some parameters degeneracies, this improves to $N_{\rm eff} = 3.30^{+0.54}_{-0.51}$ (95% CL). However in the ΛCDM model, the Planck inferred value for the Hubble constant is in tension with the local measurements, exacerbating the effect seen with pre-Planck data. The high value of local H_0 determinations could be reconciled with the CMB by going beyond the ΛCDM model and converging on values of $N_{\rm eff}$ close to 4. A detailed discussion can be found in e.g., [16].

The Planck data, within the ΛCDM framework are however also in tension with (some) measurements of the amplitude of clustering at linear-to-mildly non-linear scales. These are: the Planck Sunyaev-Zeldovich cluster abundances [17], some X-ray clusters abundances e.g., [18] and the (small scales) matter power spectrum as measured by weak lensing surveys [19, 20].

The clustering amplitude measured by these probes is lower than the one inferred from CMB data assuming a standard ($\sum m_\nu \gtrsim 0.06$ eV) ΛCDM model, a trend that would happen if neutrinos were actually massive. Note that increasing the neutrino mass above about 0.06 eV makes the tension between Planck data and local measurements of the Hubble parameter worst (which can be alleviated but at the cost of by introducing yet another parameter, i.e. allowing more than three neutrino families).

Several papers have appeared over the past few months claiming that a model where there are more than three neutrino species and neutrinos have a significantly non-zero mass, can reduce these tensions and accommodate all measurements [8, 9, 12]. Of course one cannot understate the extra-ordinary importance of the new physics that these results imply. However before interpreting this as evidence of new physics some extra care is needed.

One must first exclude the possibility that systematic biases in the measurements drive the result. New physics in the neutrino sector, therefore, is an acceptable solution if the extra parameters eliminate the tension between data sets that is present in the standard cosmological model. Moreover the trend should be seen consistently and robustly across a variety of data sets.

To be more precise, the fact that in an extended model the best fit values for extra parameters take non-standard values does not necessarily mean that the extended model provides a much improved fit to the data and that therefore should be favoured over the simpler one. This becomes an issue of model selection rather than simple parameters fitting. In the Bayesian framework (which is what is normally adopted in cosmology) model selection the Evidence ratio (or Bayes factor), gives the relative odds of two models correctly describing the observations, under the assumption of equal a priori model probabilities.

We have recently found that the extra parameters do not eliminate the tension between data sets that is present in the standard cosmological model [21].

First, models with total neutrino masses of about 0.4 eV, while may be preferred when performing a joint analysis with Planck Sunyaev-Zeldovich cluster and X-ray cluster abundances are disfavored by CMB +BAO data. Such large neutrino mass makes the tension between the

CMB-inferred H_0 and the locally measured one, very uncomfortable, pushing $N_{\rm eff}$ to high values. Therefore, the resulting constraints do not indicate a new concordance, but rather, a compromise solution between discrepant datasets.

Second, not all datasets that measure the growth of structure e.g., cluster abundance or the clustering amplitude on mildly non-linear scales agree, and the evidence for non-standard neutrino physics is actually disfavoured by other data set combinations.

Finally the evidence calculation for no data combination prefers the extended model in any significant way and for most dataset combinations actually prefers the simpler standard ΛCDM model with standard neutrinos.

6. Outlook towards the future

In the future, planned or proposed surveys will map a sizeable fraction of the observable universe. Considering only statistical errors this will open the possibility not only to measure the absolute neutrino mass scale but also to attempt to determine the mass hirerachy [22]. Neutrino hierarchy affect the shape of the matter power spectrum: neutrinos of different masses have different transition redshifts from relativistic to non-relativistic behaviour, and their individual masses and their mass splitting change the details of the radiation-domination to matter- domination regime.

Figure 2 (from [3]), shows the effect of the hierarchy on the matter power spectrum (all another quantities are kept fixed including the total neutrino mass). Also here we see that non-linearities enhance the linear theory predictions. Note that the effect is small (below the %) and that at $k \sim 1$ baryonic effects that are not included in our simulations might be at play in the Cosmos. While in principle the statistical errors from future surveys might be small enough to measure this effect, the lesson learned from §5 indicate that limits will be set by the level of control of systematics. The hierarchy signal, however small, might actually offer a very useful systematic test.

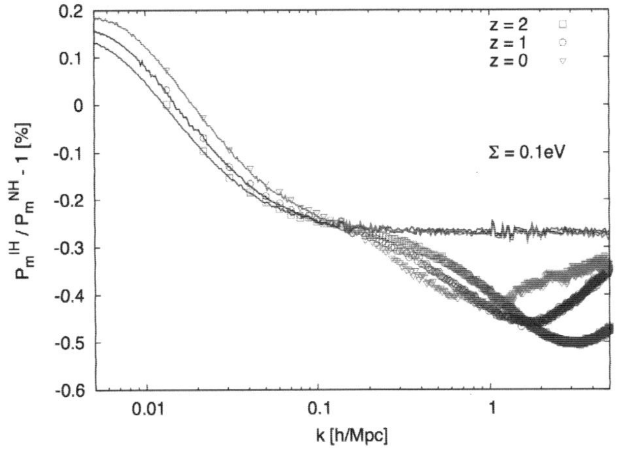

Figure 2. Fractional difference in the total matter power spectrum of the inverted hierarchy (IH) and normal hierarchy (NH) runs (we keep fixed $\sum m_\nu = 0.1$). Different colours correspond to different redshifts: green $z = 2$, blue $z = 1$, and red $z = 0$. Symbols denote the simulations results and solid lines correspond to linear theory predictions. Note that also in this case, non-linearities enhance the effect on mildly non-linear scales compared to linear theory predictions.

In order to understand this it is useful to introduce the following quantities. Let us ignore the small mass splitting and identify two mass states m the lower mass state and M the larger mass state. Then we have

$$\text{NH}: \quad \Sigma = 2m + M \quad \Delta = (M - m)/\Sigma \tag{1}$$

$$\text{IH}: \quad \Sigma = m + 2M \quad \Delta = (m - M)/\Sigma\,. \tag{2}$$

(3)

The left panel of Fig. 3 shows the regions in the Σ-Δ space that are allowed by oscillations experiments. Δ is positive for NH and negative for IH. The right panel shows the sensitivity of the matter power spectrum to Δ: clearly cosmology is mostly sensitive to the absolute value $|\Delta|$ and much less sensitive to its sign. Both figure panels are from Ref. [22]. Recall that cosmology is extremely sensitive to Σ, then to $|\Delta|$ and finally to a lesser extent to the sign of Δ. We estimate that a survey covering a sizeable part of the observable Universe and exquisite control of systematic errors could determine the hierarchy [22]; the requirements are much less stringent to constrain $|\Delta|$. Of course the sign is what gives the hierarchy but not all values of $|\Delta|$ are allowed for a given value of Σ. This offers a very powerful consistency check: when a survey measures to high statistical significance a non-zero total neutrino mass, the $|\Delta|$ value measured from the same survey should be consistent with that predicted by oscillations. Should this not be the case, then it would be a clear indication that the mass detection is affected by systematic errors.

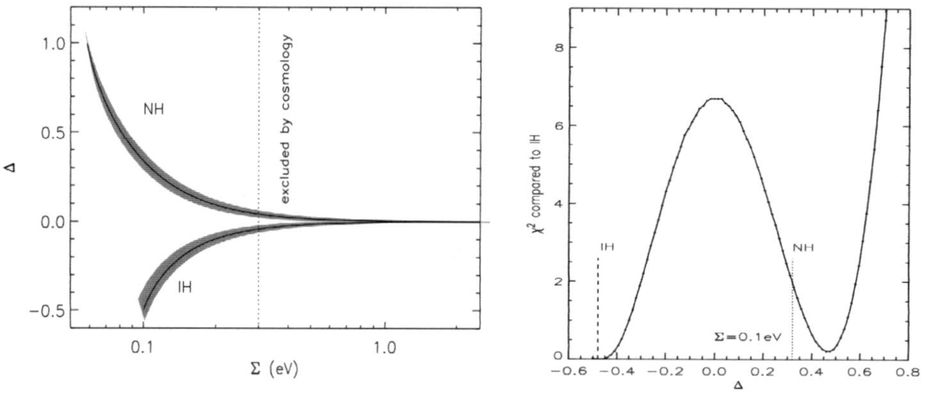

Figure 3. Figure from [22]. Left panel: regions in the Σ-Δ space that are allowed by oscillations experiments. Current cosmology limits on the total neutrino mass are also reported. Right panel: cosmology sensitivity to Δ for an idealised survey that covers a sizeable fraction of the observable Universe. Note that cosmology is much more sensitive to $|\Delta|$ than to its sign and its sensitivity to the hierarchy is due to the fact that for a given Σ $|\Delta|$ for NH is not identical to that of the IH.

7. Conclusions

If cosmology could measure not only neutrino masses but also the hierarchy, in combination with neutrino-less double beta decay experiments, this would help answering the question of wether neutrinos are Majorana or Dirac particles. This is yet another example of how the deep connections and how powerful is the interplay between the infinitely big and the infinitely small.

Precision cosmology means that we can start (or prepare for) constraining interesting physical quantities by looking up at the Cosmos. Here I have concentrated on neutrino properties, absolute mass scale, number of species and possibly the mass hierarchy.

Recent contradictory statements have appeared in the literature: neutrino mass detections that are in tension with 95% upper limits and detection of extra species which are in tension with constraints on deviations from three families. My personal view is that once the dust has settled, we will converge to a constrain of $\sum m_\nu \lesssim 0.25$ eV (95% CL) and consistency with 3 families.

Large future surveys means that sub % effects become detectable, which brings in a whole new set of challenges but also of opportunities, examples that I have discussed are the absolute neutrino mass scale, the mass hierarchy. The (indirect) detection of neutrino masses from cosmology is within the reach of forthcoming experiments (even for the minimum mass allowed by oscillations). However systematic and real-world effects are the challenge. A lot of work is needed in this direction and there is a clear need for in-built consistency checks. There are exciting times ahead.

References

[1] Planck Collaboration, Ade, P. A. R., Aghanim, N., et al. 2013, arXiv:1303.5076
[2] Carbone, C., Verde, L., Wang, Y., & Cimatti, A. 2011, JCAP, 3, 30
[3] Wagner, C., Verde, L., & Jimenez, R. 2012, ApJLett, 752, L31
[4] Bird, S., Viel, M., & Haehnelt, M. G. 2012, MNRAS, 420, 2551
[5] Smith, K. M., Zahn, O., & Doré, O. 2007, PRD, 76, 043510
[6] Planck Collaboration, Ade, P. A. R., Aghanim, N., et al. 2013, arXiv:1303.5077
[7] The POLARBEAR Collaboration, Ade, P. A. R., Akiba, Y., et al. 2014, arXiv:1403.2369
[8] Battye, R. A., & Moss, A. 2014, Physical Review Letters, 112, 051303
[9] Wyman, M., Rudd, D. H., Vanderveld, R. A., & Hu, W. 2014, Physical Review Letters, 112, 051302
[10] Hamann, J., & Hasenkamp, J. 2013, JCAP, 10, 44
[11] Giusarma, E., Di Valentino, E., Lattanzi, M., Melchiorri, A., & Mena, O. 2014, arXiv:1403.4852
[12] Dvorkin, C., Wyman, M., Rudd, D. H., & Hu, W. 2014, arXiv:1403.8049
[13] Archidiacono, M., Giusarma, E., Melchiorri, A., & Mena, O. 2012, PRD, 86, 043509
[14] Riemer-Sørensen, S., Parkinson, D., Davis, T. M., & Blake, C. 2013, ApJ, 763, 89
[15] Feeney, S. M., Peiris, H. V., & Verde, L. 2013, JCAP, 4, 36
[16] Verde, L., Protopapas, P., & Jimenez, R. 2013, Physics of the Dark Universe, 2, 166
[17] Planck Collaboration, Ade, P. A. R., Aghanim, N., et al. 2013, arXiv:1303.5080
[18] Vikhlinin, A., Kravtsov, A. V., Burenin, R. A., et al. 2009, ApJ, 692, 1060
[19] Kilbinger, M., Fu, L., Heymans, C., et al. 2013, MNRAS, 430, 2200
[20] Heymans, C., Grocutt, E., Heavens, A., et al. 2013, MNRAS, 432, 2433
[21] Leistedt B., Peiris, H. V., Verde, L., to be sumbitted.
[22] Jimenez, R., Kitching, T., Peña-Garay, C., & Verde, L. 2010, JCAP, 5, 35

Acknowledgments

European Research Council under the European Community's Seventh Framework Programme FP7-IDEAS-Phys.LSS 240117 and Mineco grant FPA2011-29678-C02-02. I would like to thank my collaborators in this work: Carmelita Carbone, Stephen Feeney, Raul Jimenez, Boris Leistedt, Hiranya Peiris, Christian Wagner.

Journal of Physics: Conference Series **598** (2015) 012011 doi:10.1088/1742-6596/598/1/012011

MEMPHYS: A next generation megaton water Cherenkov detector in Europe

Luca Agostino on the behalf of the MEMPHYS collaboration
AstroParticule et Cosmologie (APC), CNRS
E-mail: luca.agostino@apc.univ-paris7.fr

Abstract. MEMPHYS (MEgaton Mass PHYSics) is a proposed large-scale water Cherenkov detector to be installed deep underground. It is dedicated to nucleon decay searches, neutrinos from supernovae, solar and atmospheric neutrinos, as well as neutrinos from a future Super-Beam or Beta-Beam to measure the CP violating phase in the leptonic sector and the mass hierarchy with atmospheric neutrinos. This proceeding will show the status of the design studies concerning in particular the optical layout and the dedicated test-bench (MEMPHYNO). In addition, some results in terms of physics performances obtained by simulations will be shown.

1. Introduction

A megaton-scale water Cherenkov detector would have competitive capabilities for accelerator-based neutrino oscillation physics such as the possibility to discover the CP violation in the leptonic sector. In addition, it would reach a sensitivity on the proton lifetime close to the predictions of most supersymmetric or higher dimension grand unified theories and it would explore neutrinos from supernovae and from other astrophysical sources.

The existing Water Cherenkov Detectors are a precious source of expertise and information for MEMPHYS, which will be roughly 20 times the size of SuperKamiokande [1], so MEMPHYS will be a reasonable extension of a known, well performing detector.

2. The MEMPHYS (Megaton Mass PHYSics) detector

MEMPHYS [2] [3] is a proposed large-scale water Cherenkov detector with a fiducial mass of about 500 kt. It is composed by two tanks of 65m in diameter and 103m in height.

Each module is equipped with ~120000 12" photomultipliers (PMTs) providing 30% optical coverage. The detector could be installed at the Fréjus site, near the existing Laboratoire Souterrain de Modane (LSM laboratory) located at 130 km from CERN and with a rock overburden of 4800 m.w.e.

3. Optical layout optimization

Optimization studies to reduce the number of PMTs have been done investigating different possibilities: comparison of different photocathode diameters (8", 10" and 12"), the using of High Quantum Efficiency (HQE: 32% at 400 nm) and Normal Quantum Efficiency (NQE: 22% at 400 nm) photocathodes, different geometrical coverages (30%, 20% and 15%) and the possibility of adding Light Concentrators (LC).

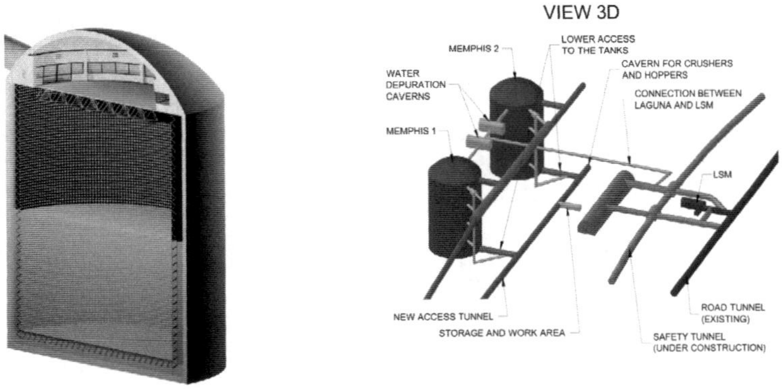

Figure 1. Schematic view of one MEMPHYS module (left) and design for installation and infrastructure at a possible extension of the LSM undeground laboratory at the Fréjus site (right).

As an example of the design optimization tests, in Figure 2 we show the mean number of collected PEs per MeV and the mean number of hit PMTs as a function of the electron energy considering 12" PMTs and different geometrical coverages, NQE or HQE, with or without LC.

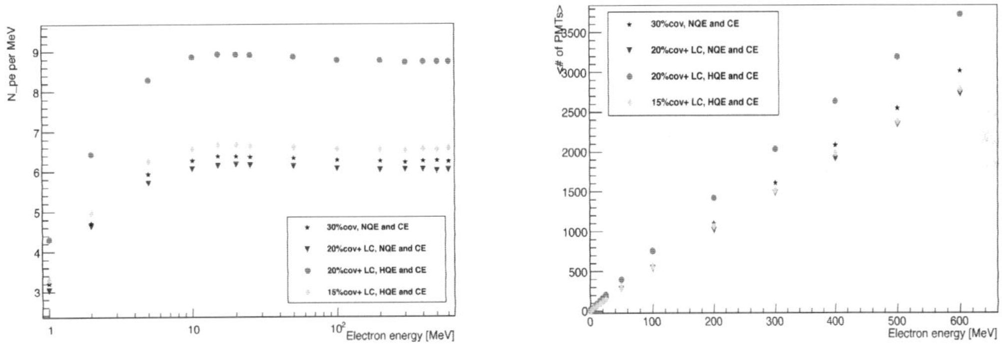

Figure 2. Mean number of collected photoelectrons per MeV (left) and mean number of hit PMTs (right) as a function of the electron energy for different configurations. The CE is set to 0.8 and considered PMTs are 12".

4. MEMPHYNO test-bench

One important R&D item towards the construction of MEMPHYS and other large liquid-based detectors is focused on the reduction of the number of electronics channels for power supply and signal read-out of the PMTs, which is expected to be one of the major costs of the experiment. The PMm^2 R&D programme [4] has developed an integrated readout electronics circuit (an ASIC called PARISROC [5]) for groups of PMTs (matrix of 4x4). MEMPHYNO [6] is a test bench for light sensor or electronics solutions installed at the APC Laboratory in Paris. It consists of a high-density polyethylene tank of 2x2x2 m^3, filled with water and placed inside a muon hodoscope. A photograph of the detector and of the PMT matrix as well as an example of signal are shown in Figure 3.

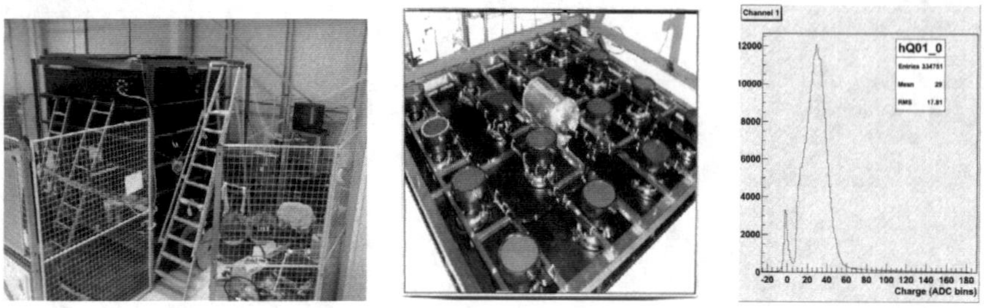

Figure 3. Picture of the MEMPHYNO test-bench (left) and of the PMT matrix (center) and an example of signal recoiled on one PMT (right).

5. Simulation and Physics Studies

Realistic analysis algorithms have been developed to evaluate the detector efficiency and energy resolution, which are then used for studies on the MEMPHYS physics reach using the GLoBES package. [7].

In order to properly take into account all the effects of the reconstruction, the detector performance is conventionally described in terms of "Migration Matrices" representing the neutrino reconstructed energy versus the true one.

An example of these matrices is shown in Figure 4 (left). In the right panel of Figure 4 the coverage of the CP phase space at 3σ is shown as a function of the true value of $\sin^2 2\theta_{13}$ for different values of the systematic errors. The beam consider is the Super Beam [8] from CERN studied in the context of the EURONU project [9] We expect to cover about 55% at 3σ with 5%/10% error on signal and background.

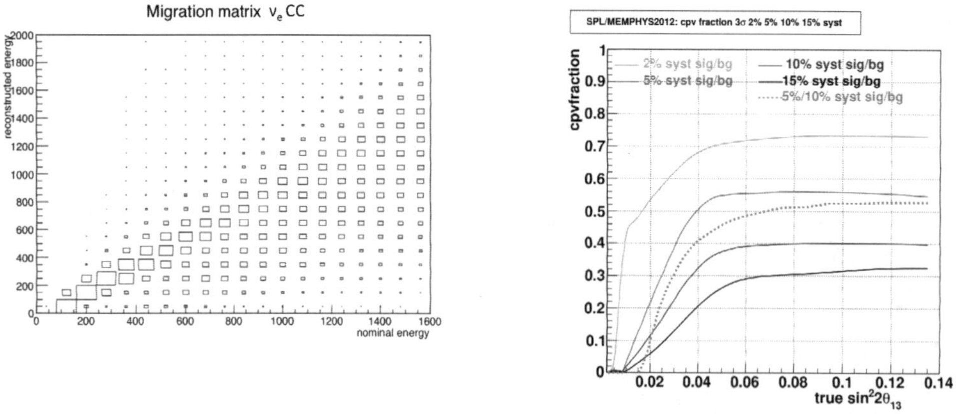

Figure 4. "Migration Matrices" with reconstructed neutrino energy as a function of true energy for selected events (ν_e CC)

[1] M. Koshiba et al., Nuovo Cim. C9 (1986) 141-158

[2] L. Agostino *et al.* [MEMPHYS Collaboration], JCAP **1301** (2013) 024 [arXiv:1206.6665[hep-ex]].
[3] L. Agostino *et al.* [MEMPHYS Collaboration], Phys. Rev. ST Accel. Beams **16** (2013) 061001 .
[4] B. Genolini et al. Nucl.Instrum.Meth, A610 (2009) 249-252
[5] G. Martin-Chassard et al. Nucl.Instrum.Meth.", A623 (2010) 492-494
[6] M. Marafini, PhD Thesis, Universit Paris Diderot, Paris, France (2011)
[7] P. Huber et al., Comput. Phys. Commun. 167 (2005) 195 and Comput. Phys. Commun. 177:432-438, 2007
[8] A. Longhin, Eur.Phys.J. C71 (2011) 1745
[9] T.R. Edgecock et al. Phys. Rev. ST Accel. Beams 16 021002 (2013)

Journal of Physics: Conference Series **598** (2015) 012012 doi:10.1088/1742-6596/598/1/012012

Results from EXO–200

D.J. Auty on behalf of the EXO–200 Collaboration

Dep. of Physics and Astronomy,University of Alabama,206 Gallalee Hall, Tuscaloosa, AL 35487-0324,USA

E-mail: dauty@ua.edu

Abstract. EXO–200 is a double-beta decay experiment that measures the half-life of the decay of ^{136}Xe. This article describes the result reported by EXO–200 in the summer 2012, with an exposure of 32.5 kg·yr, saw no signal for the 0 $\nu\beta\beta$ decay. This sets a lower limit for the half-life of the neutrinoless double-beta decay $T^{0\nu\beta\beta}_{1/2}$ (^{136}Xe)> 1.6×10^{25}yr(90 % CL), which sets a limit on the Majorana neutrino mass, depending on which matrix element used, of 140 meV–380 meV.

1. Introduction

Several even-even nuclei are stable against ordinary β decay, but are unstable for $\beta\beta$ decay in which two neutrons are changed into two protons simultaneously. As is well known, $\beta\beta$ decay can proceed through several modes. The allowed process, the two-neutrino mode ($2\nu\beta\beta$), is completely described by known physics; its rate was first evaluated in [1]. Of the other, hypothetical, modes, the neutrino-less decay ($0\nu\beta\beta$) is forbidden in the Standard Model since it violates conservation of the total lepton number. EXO–200 is located at the Waste Isolation Pilot Plant (WIPP) in New Mexico, USA. EXO–200 uses xenon as both the $\beta\beta$ source and the detector of the summed electron energy. It uses 200 kg, of which 175 kg is in liquid phase, of xenon enriched to (80.6±0.1) % in the isotope ^{136}Xe. The remaining 19.4 % is ^{134}Xe, with other isotopes present only at low concentration. At operating temperature, 167 K, and a pressure, 147 kPa, the liquid xenon (LXe) has a density of 3.0 g cm^{-3}.

EXO-200 has been described in detail elsewhere [2]. Briefly, the detector is a cylindrical LXe time projection chamber (TPC), roughly 40 cm in diameter and 44 cm in length. Two drift regions are separated in the centre by a cathode. The TPC provides X-Y-Z coordinate and energy measurements of ionisation deposits in the LXe by simultaneously collecting the scintillation light and the charge. Charge deposits spatially separated by about 1 cm or more are individually observed and the position accuracy for isolated deposits is a few mm. Avalanche Photodiodes (APDs) measure the scintillation light. Small radioactive sources can be positioned at standard positions near the TPC to calibrate the detector and monitor its stability. The TPC is shielded from environmental radioactivity on all sides by ∼50 cm of HFE-7000 cryofluid [3] maintained at ∼167 K inside a vacuum-insulated copper cryostat. Further shielding is provided by at least 25 cm of lead in all directions. The entire assembly is housed underground in a cleanroom at a depth of 1585+11 meters water equivalent [4] at the Waste Isolation Pilot Plant near Carlsbad, NM, USA. Four of the six sides of the cleanroom are instrumented with plastic scintillator panels recording the passage of cosmic ray muons. An extensive materials screening campaign [5] was employed to minimise the radioactive background produced by the detector components.

2. Data

The active volume of the LXe for this analysis was the inner 98.5 kg LXe, 79.4 kg of ^{136}Xe. The data analysed was taken between 22nd September 2011 and 16th April 2012, for a total of 2896.6 live hours after cuts [6], which is 32.5 kg·yr livetime. Only events above 700 keV are included in this analysis, as this is where the trigger is 100 % efficient. Other analysis cuts that were used were: events in the TPC that happened 1 μs before or 25 ms after a veto panel triggered were rejected, making the veto efficiency $96.2^{+0.4}_{-3.7}$ % at rejecting cosmic muons (0.58 % dead time); TPC events that happened within one second of each other to be rejected (3.3 % dead time), TPC events that happened within one minute after a muon was reconstructed in the TPC (5.0 % dead time), to reduce the daughter decays following the muon; alpha decays have a larger light to charge ratio than betas so were able to be cut on this. The total dead time due to the veto cuts was 8.6 %. The stability of the detector was achieved through daily calibrations mostly using the ^{228}Th source (2615 keV). A ^{60}Co (1173 keV and 1332 keV) and ^{137}Cs (662 keV) source were used less often to calibrate the energy for the full spectrum. The energy from an event can go into light and charge by various amounts, so smearing any peak out in either spectrum. We are able to use the known energy of the 2615 keV peak from ^{208}Tl decay to reduce the resolution by combine the light and charge by "rotating", $\mathrm{E} = \mathrm{S}\sin\theta + \mathrm{I}\cos\theta$, and fitting a Gaussian to the peak. Theta was varied to get the lowest width/mean for the fit. This improve the energy resolution compared to scintillation or charge only. The whole of the low background data was rotated with this theta, and the energy resolution parameterised as $\sigma^2 = a\sigma_e^2 + bE + E^2$ where σ_e^2 is the electronic noise contribution, bE is the statistical fluctuation in the ionisation and scintillation and cE^2 is the position and time dependant broadening. In the rotated energy spectrum at the Q-value the energy resolution was 1.67 % (1.84 %) SS (MS).

3. Analysis

The detectors ability to identify singlesite charge clusters (SS) and multi-site charge clusters (MS) events to distinguish between SS $\beta/\beta\beta$ events and MS γ events in the bulk xenon. The clustering resolution in two dimensions, U-dimention is 18 cm and 6 mm in z (drift time). The ability of the GEANT4 Monte Carlo (MC) to simulate the distributions for SS and MS energy spectra was proven on the ^{228}Th and ^{60}Co source data. The MC reproduces the fraction of SS events, defined as $N_{ss}/(N_{ss} + N_{ms})$ to ± 8.5 %. The requirement that events be fully reconstructed[1] gives an efficiency of 70 % (71 %) for the ^{228}Th ($0\nu\beta\beta$) spectrum. The 71 % efficiency was further verified by comparing the $2\nu\beta\beta$ data spectrum with MC. ground Background models were generated for the various components of the detector from the material screening campaign, providing normalisation for the MC generated probability distribution functions (pdfs). These background contributions were estimated using previous generation of detector simulations [2], pdfs which account for <50 (<0.2) events in the $2\nu\beta\beta$ ($0\nu\beta\beta$) spectrum, were not included in the fit. The various generated pdfs were fitted to the data using a maximum likelihood fit and varied so that SS site events were constrained to be $\pm$8.5 % of the value predicted by the MC. As γ-rays were used to calibrate the detector, the energy of the β events was a free parameter in the fit, so it is constrained by the $2\nu\beta\beta$ events. The fit reports a scale factor of 0.995 $\pm$0.004. The fit gives a expected background of $(1.5\pm0.1)\times10^{-3}\,\mathrm{kg}^{-1}\,\mathrm{yr}^{-1}\,\mathrm{keV}^{-1}$ in the one sigma region of interest (ROI), where one sigma is 1.67 % of 2458 keV(41 keV), which is 3–4 events. Figure 1 shows that there is only one event in this region. The result from the fit is a $\mathrm{T}_{1/2}^{0\beta\beta} > 1.6\times10^{25}$ yr with the $\mathrm{T}_{1/2}^{2\beta\beta} = 2.23\pm0.02$ (stat.)$\pm$0.22 (syst.)$\times10^{21}$yr. The dominate systematic errors are from the fiducial volume (12.34 %) and the β scale (9.32 %). $\mathrm{T}_{1/2}^{0\beta\beta}$ can be converted into a limit on neutrino mass using various matrix elements which give a neutrino mass between 140 meV and 380 meV at 90 % CL. In Figure 2 the comparison between $\mathrm{T}_{1/2}^{0\beta\beta}$ for

[1] have charge and scintillation signal

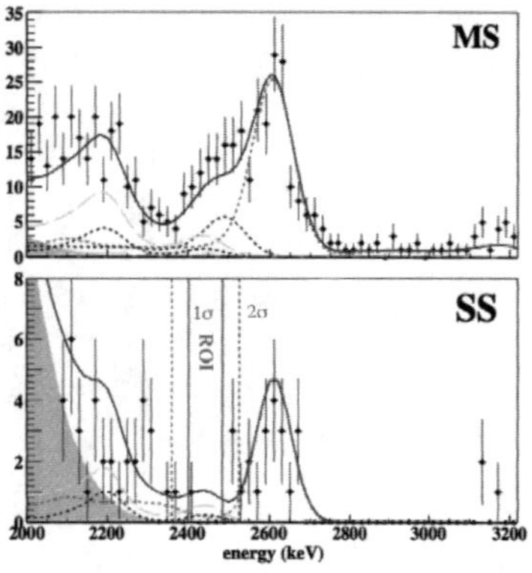

Figure 1. Energy spectra in the ^{136}Xe $Q_{\beta\beta}$ region for MS (top) and SS (bottom) events. The 1 (2)σ regions around $Q_{\beta\beta}$ are shown by solid (dashed) vertical lines. The best fit line (solid blue) is shown. The background components are $2\nu\beta\beta$ (grey region), ^{60}Co (dotted dark blue), ^{222}Rn in the cryostat-lead air-gap (long- dashed green), ^{238}U in the TPC vessel (dotted black), ^{232}Th in the TPC vessel (dotted magenta), ^{214}Bi on the cathode (long-dashed cyan),The $0\nu\beta\beta$ PDF from the fit is not visible.

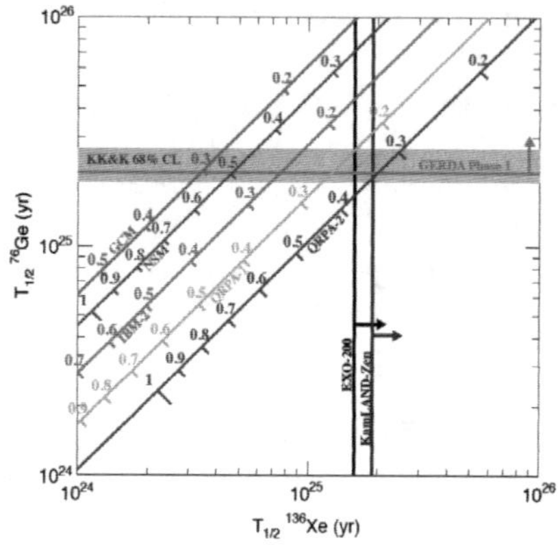

Figure 2. Relation between the $T_{1/2}^{0\nu\beta\beta}$ in ^{76}Ge and ^{136}Xe for different matrix element calculations (GCM [8], NSM [9], IBM-2 [10], RQRPA-1 [11] and QRPA-2 [12]). For each matrix element $\langle m\rangle_{\beta\beta}$ is also shown (eV). The claim [13] is represented by the grey band, along with the best limit for ^{76}Ge [14]. The result reported here is shown along with that from [7].

^{76}Ge and ^{136}Xe is shown with the predicted Majorana mass of the neutrino, also plotted are the limits set by this analysis and a previous analysis for KamLAND-Zen [7] for ^{136}Xe and the claimed observation and the 90 % limit for ^{76}Ge. It can be seen that for most matrix elements the 90 % limit from EXO is incompatible with the Ge observation. A new result released from EXO–200, after this poster was presented, using 99.8 kg·yr of ^{136}Xe has given a 90 % CL $T_{1/2}^{0\nu} > 1.1 \cdot 10^{25}$yr [15].

Acknowledgements

EXO–200 is supported by DoE and NSF in the United States, NSERC in Canada, SNF in Switzerland and RFBR in Russia. The collaboration gratefully acknowledges the WIPP for their hospitality

References

[1] M. Goeppert-Mayer, *Phys. Rev.* **48**, 512 (1935).
[2] M. Auger *et al.*, *JINST* **7**, P05010 (2012).
[3] 3M HFE-7000, http://www.3m.com/.
[4] E.I. Esch *et al.*, *Nucl. Instrum. Methods* A **538**, 516 (2005).
[5] D.S. Leonard *et al.*, *Nucl. Instrum. Methods* A **591**, 490 (2008).
[6] M. Auger *et al.*, *Phys. Rev. Lett.* **109**, 032505 (2012).
[7] A. Gando *et al.*, *Phys. Rev. Lett.* **110**, 062502 (2013).
[8] T.R. Rodriguez and G. Martinez-Pinedo, *Phys. Rev. Lett.* **105**, 252503 (2010).
[9] J. Menendez *et al.*, *Nucl. Phys.* A **818**, 139 (2009).

[10] J. Barea and F. Iachello, *Phys. Rev.* C **79**, 044301 (2009).
[11] F. Simkovic *et al.*, *Phys. Rev.* C **79**, 055501 (2009).
[12] A. Staudt, K. Muto and H. V. Klapdor-Kleingrothaus, *Europhys. Lett* **13**, 31 (1990).
[13] H.V. Klapdor-Kleingrothaus and I.V. Krivosheina, *Mod. Phys. Lett.* A **21**, 1547 (2006).
[14] M. Agostini *et al. Phys. Rev. Lett.* **111**, 122503 (2013)
[15] J.B. Albert *et al.* arXiv:1402.6956 [nucl-ex].

Current status of the SNO+ detector: preparations for first physics data

A R Back for the SNO+ Collaboration

School of Physics and Astronomy, Queen Mary University of London, 327 Mile End Road, London, E1 4NS, UK

E-mail: a.r.back@qmul.ac.uk

Abstract. SNO+ is a multi-purpose neutrino detector, which will primarily study neutrinoless double beta decay. Preparations are underway for the first phase of physics data-taking, with a water-filled detector. In terms of hardware, recent electronics repairs are described, as well as the programme which has shown to be ~90% effective in repairing failed photomultiplier tubes. Testing of Data Quality has revealed some areas in the software that could be developed further and has provided useful cross-checks of trigger behaviour.

1. Introduction

The SNO+ detector [1] is located at SNOLAB, a Canadian laboratory two kilometres underground (6000 mwe), in Creighton mine, near Sudbury, ON. Its key physics goal is to probe neutrinoless double beta decay with the isotope ^{130}Te. The varied physics program also hopes to build on the success of predecessor the Sudbury Neutrino Observatory (SNO) [2], in solar and supernova neutrinos, among others. To achieve these goals the detector has been upgraded, so that it can be filled with nearly one kilotonne of liquid scintillator; first scintillator data is expected in autumn 2015. Considerable upgrades are required to transform a D_2O detector, sensitive to Čerenkov light, into a liquid scintillator detector, with 50 times higher light yield. For example a hold-down rope-net has been installed [3] to combat the buoyancy of a scintillator-filled acrylic vessel (AV).

In these proceedings, Section 2 provides details on some of the more recent detector work. The cavity outside the AV is currently being filled with ultrapure water. In February and March 2014 commissioning detector running took place, with ~100 photomultiplier tubes (PMTs) submerged. Section 3 focusses on recent developments in software, namely SNO+ Data Quality, which have used these commissioning runs for testing, in preparation for first physics data-taking.

2. Hardware preparations

The electronics that handle triggering and data acquisition (DAQ), were inherited from SNO, but have been upgraded, in places, to cope with the higher data rate expected for SNO+. Testing during two short periods of (air-filled) detector running, in February and October 2012, revealed over 200 (of about 9600) electronics channels not functioning as expected. In the majority of cases the problem was with the board that prepares the raw PMT signal before it enters the trigger electronics—the PMT Interface Card (PMTIC). Simple tests of the resistance and capacitance at strategic points on the PMTIC, usually identified a blown resistor or capacitor as the problem, which was then promptly replaced. The electronics repairs performed in autumn 2013, saw around 90% of these bad channels inspected and successfully repaired, as confirmed by the recent commissioning runs.

Like the AV and triggering/DAQ electronics, SNO+ has inherited its PMTs from SNO. SNO was designed to have close to 10,000 Hamamatsu R1408 PMTs (see [2] for a detailed schematic), but about 800 PMTs failed during its lifetime. Some were removed to accommodate the hold-up rope-net, yet SNO+ is still designed to have nearly 9600 PMTs. Since increasing the PMT coverage of the AV directly improves energy resolution, some of the failed ones are being repaired. So far about 300 PMTs have been repaired at Laurentian University (and previously at Queens University). The remaining failed PMTs will be accessed by boat during the water-fill, so a quick turnaround on removing, repairing and replacing them is required.

The remainder of this section aims to provide a brief overview of the PMT repair process. Work begins with some basic tests using the dark box set-up pictured in Figure 1(a), predominantly to confirm that the PMT actually needs to be repaired and to begin isolating the problem. The resistance and capacitance are measured and the potential across the PMT is gradually ramped up to its operating voltage; to check for breakdowns. In most cases signs of failure will be visible by this point, but if not the trigger rate at different thresholds is also tested. Repair work so far has found that in around 90% of cases the PMT works as expected after replacing the base.

(a) (b) (c)

Figure 1. (a) – The full set-up used for dark box testing. (b) – The lower part of a PMT being cleaned in mineral solution. (c) – The hub being potted with the two-part SilGel mix.

In order to replace the base, the PMT must be disassembled and cleaned. It is dismantled by meticulously slicing through the heat shrink wrap, being careful not to damage the hub, which is re-used, or PMT. Once the heat shrink wrap is pealed off, the hub must be carefully prised away from the silicone gel—used to create a watertight seal and protect the base and pins. The base is de-soldered and discarded and the majority of the gel is removed, before soaking in mineral spirits (Figure 1(b)) to break down any remaining gel. Once cleaned thoroughly, a new base is soldered on and the dark box testing is repeated to verify that it is operating as normal.

In the final stage the original hub is affixed using a new heat shrink wrap and glue tape, which are both heated steadily and evenly for a good seal. The seal is further improved by applying a conformal coating to the inside of the hub and electronics. The potting process sees the hubs refilled with SilGel, a two-part silicon mix, using a nozzle that ensures the two components are correctly mixed (Figure 1(c)). The SilGel shrinks as it sets so the hubs are topped up before being capped. The PMT is ready to be installed back in the detector once the cap has been sealed in place.

Journal of Physics: Conference Series **598** (2015) 012013 doi:10.1088/1742-6596/598/1/012013

3. Data Quality software

As well as hardware, the recent commissioning runs have seen equally significant developments in software areas. One such area concerns the quality control and selection of runs for physics analyses, which is critical for SNO+ to achieve its physics goals [1, 3]. Although, a varied and comprehensive set of online monitoring tools will be available, the offline SNO+ Data Quality software has the opportunity to consider entire runs, as well as the behaviour of the detector as a whole. The software was developed by first considering the run selection checks used in SNO [4], and forming groups of similar checks. Two of these groups—including the most recently developed group of timing checks, which examines the time interval between consecutive events (Δt)—were tested on recent commissioning data. The performance of all the checks was measured across 42 runs in the February stable running period. Whilst some checks performed as expected, others failed on almost all runs, highlighting that further development is required. These issues are being addressed and recent reprocessing of the data revealed a significant improvement in performance in one of the checks.

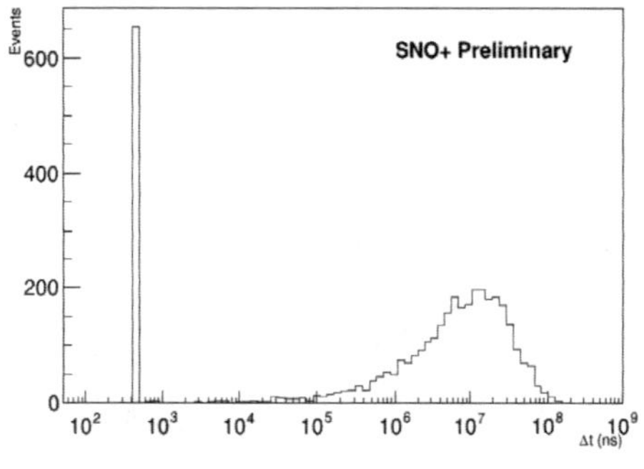

Figure 2. Δt distribution from one particular run during the February 2014 stable running period. Note the large peak at ∼400 ns.

SNO+ uses a 400 ns trigger window in which information is collected from every PMT that fired. The master trigger card has the functionality to initiate a new 400 ns trigger window, directly following the current window, with 30–50 ns down time separating the two. This is referred to as an automatic re-trigger. Figure 2 provides an example Δt distribution, where a peak at around 400 ns is evident. Further analysis found it to be at 440 ns, which corresponds well with the signature of an automatic re-trigger. Detector operators confirmed that automatic re-triggers were enabled for every event, in a particular trigger channel, so this analysis has proved a valuable cross-check.

4. Conclusion

The PMT repair programme has already proved successful in repairing ∼90% of failed PMTs, and will improve the energy resolution of the detector. Software frameworks, such as SNO+ Data Quality have effectively used the recent commissioning data as a platform for testing, whilst it has also been used to validate recent electronics repairs. These are important steps towards readiness for physics data taking and our first physics results.

References

[1] Chen M C 2005 *Nucl. Phys. B—Proc. Sup.* **145** 65
[2] The SNO Collaboration 2000 *Nucl. Instrum. Meth. A* **449** 172
[3] Lozza V 2014 *EPJ Web of Conferences* **65** 1003
[4] Wilson J 2004 Ph.D. thesis University of Oxford

Precision measurements of θ_{12} for testing models of discrete leptonic flavour symmetries

P. Ballett[1†], S. F. King[2], C. Luhn[3], S. Pascoli[1] and M. A. Schmidt[4]

[1]IPPP, Department of Physics, Durham University, South Road, Durham DH1 3LE, United Kingdom

[2]School of Physics and Astronomy, University of Southampton, Highfield, Southampton SO17 1BJ, United Kingdom

[3]Theoretische Physik 1, Naturwissenschaftlich-Technische Fakultät,Universität Siegen, Walter-Flex-Straße 3, D-57068 Siegen, Germany

[4]ARC Centre of Excellence for Particle Physics at the Terascale, School of Physics, The University of Melbourne, Victoria 3010, Australia

Abstract. Models of leptonic flavour with discrete symmetries can provide an attractive explanation of the pattern of elements found in the leptonic mixing matrix. The next generation of neutrino oscillation experiments will allow the mixing parameters to be tested to a new level of precision, crucially measuring the CP violating phase δ for the first time. In this contribution, we present results of a systematic survey of the predictions of a class of models based on residual discrete symmetries and the prospects for excluding such models at medium- and long-term oscillation experiments. We place particular emphasis on the complementary role that a future circa 50 km reactor experiment, *e.g.* JUNO, can play in constraining these models.

1. Introduction

The next generation of neutrino oscillation experiments has the potential to not only determine the remaining unknowns in the PMNS matrix but also to measure its parameters with unprecedented precision. This will mark the beginning of a period of high-precision neutrino physics, where the standard paradigms describing the neutrino sector will be put to proof and theoretical ideas about the origins of neutrino mass and leptonic flavour can be confronted with data.

One of the more popular beyond the standard model ideas applied to the neutrino sector is the introduction of a discrete flavour symmetry. Models based on this principle have been shown to be able to derive the observed structure of the PMNS matrix from a small set of assumptions. These models generally propose a discrete symmetry (*e.g.* A_4 or S_4) which is broken spontaneously, leaving residual symmetries amongst the leptonic mass terms. These symmetries reduce the degrees of freedom amongst the mixing parameters, generating a pattern of falsifiable predictions. By hypothesizing which symmetries of the leptonic mass terms are residual, this idea can be used to reconstruct the flavour group independently of many model specific assumptions. In Ref. [1], we have shown that a quite general construction of this type (first presented in Ref. [3]) leads to only 8 viable models in light of the current global oscillation data. These models fix a column of the PMNS matrix which, under the assumption of unitarity,

† Poster presenter at NuPhys 2013 held at Institute of Physics, London, UK.

Model label	solar prediction	predicted θ_{12} (r in 3σ)
A_4 T_α–S_2	$s = \sqrt{\frac{2}{2-r^2}} - 1$	[35.62, 35.86]
S_4 T_e–S_1	$s = \sqrt{1 - \frac{2r^2}{2-r^2}} - 1$	[34.05, 34.55]
S_4 T_α–S_2	$s = \sqrt{\frac{3}{2(1-r^2)}} - 1$	[30.29, 30.49]
A_5 T_e–S_1	$s = \sqrt{3 + \frac{6}{(3-\varphi)(r^2-2)}} - 1$	[30.33, 30.90]
A_5 T_e–S_2	$s = \sqrt{\frac{6}{(2+\varphi)(2-r^2)}} - 1$	[32.03, 32.24]
A_5 T_α–S_2	$s = \sqrt{\frac{3\varphi}{(2\varphi-1)(2-r^2)}} - 1$	[37.56, 37.62]

Table 1. The solar predictions for the 8 viable models identified in Ref. [1]. The model label denotes the flavour group and the pattern of breaking; for details, see Ref. [1].

can be expressed in terms of two constraints on the PMNS parameters: the *atmospheric sum rule*, relating θ_{23} to θ_{13} and δ, and the *solar prediction*, an expression for θ_{12} in terms of θ_{13} alone. In this contribution, we shall discuss the parameter correlations of these models, and how they can be constrained by the next generation of high-precision oscillation experiments. We shall employ the notation $s = \sqrt{3}\sin\theta_{12} - 1$, $r = \sqrt{2}\sin\theta_{13}$ and $a = \sqrt{2}\sin\theta_{23} - 1$ [2] throughout.

2. Atmospheric sum rules

The atmospheric sum rule can be written in a linearized form by

$$a = a_0 + \lambda r \cos\delta + \mathcal{O}(r^2),$$

where a_0 and λ are constants expressible in terms of the group theoretic parameters of each model. To test these relations, we require a strong precision on the parameter δ, which necessitates the consideration of the next generation of long-baseline experiments. These proposals seek to make accurate measurements of the appearance channels $\nu_\mu \to \nu_e$ and $\overline{\nu}_\mu \to \overline{\nu}_e$, which are sensitive to the value of δ at a subdominant level. Although it remains a challenging measurement, two leading designs have been shown to offer significant sensitivity to δ: superbeams and neutrino factories. Such facilities would be able to constrain the atmospheric sum rules over a significant fraction of the available parameter space. For example, an on-axis superbeam with a detector mass of 70 kton (35 kton) and a baseline of 2000 km would be capable of excluding models with $a_0 = 0$ and $\lambda = 1$ for over 44–88% (19–75%) of the parameter space, depending on the true value of θ_{23} [1].

3. Solar predictions and reactor experiments

In this section, we shall consider a circa 50 km reactor experiment based on the Jiangmen Underground Neutrino Observatory (JUNO) [4] and Reactor Experiment for Neutrino Oscillations (RENO-50) designs [5]. These facilities will be capable of high precision measurements of the $\overline{\nu}_e$ disappearance probability. The main goal of such experiments is to observe the subdominant oscillations whose phase depends upon the mass hierarchy. However, they will also significantly increase the precision on the oscillation parameters θ_{12}, Δm^2_{21} and Δm^2_{31}, reducing their uncertainty to the sub-percent level. The 8 viable models identified in Ref. [1] make 6 distinct solar predictions, which are shown in Table 1. Such precision will have a significant impact on the viability of the correlations predicted by flavour symmetric models. To understand the impact of these high precision measurements, we have performed a simulation based on the JUNO design to determine its ability to test the correlations shown in Table 1. In

Journal of Physics: Conference Series **598** (2015) 012014 doi:10.1088/1742-6596/598/1/012014

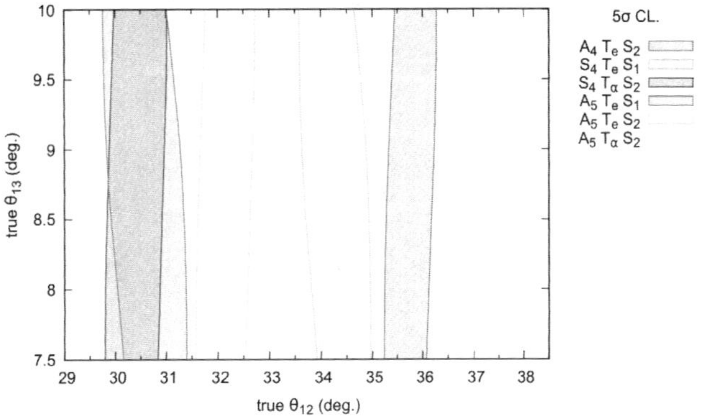

Figure 1. The 5σ allowed regions for the solar predictions shown in Table 1 after 6 years of data taking by JUNO.

our simulation, we assume a 20 kton liquid scintillator detector with a linear energy uncertainty of $0.03/\sqrt{E}$. The JUNO facility will detect neutrinos from 10 nearby reactors; however, we model this by a single source at a baseline distance given by the power weighted average of 52.5 km and a reactor power of 36 GW [4]. We have normalised our spectrum to produce 10^5 events, including a 5% normalisation uncertainty. In Fig. 1, we show the allowed regions at 5σ significance for the models shown in Table 1. We see that only two of the 5σ intervals overlap, which allows for a strong model discrimination. The ability for JUNO to exclude these models independently of their atmospheric sum rules provides a great complementarity between the reactor and long-baseline programmes. Furthermore, the two indistinguishable models for JUNO predict very different atmospheric sum rules,

$$a = \pm\frac{1}{6} - \frac{1}{\sqrt{6}} r\cos\delta \quad (\mathrm{S}_4\ \mathrm{T}_\alpha\text{–}\mathrm{S}_2) \qquad \text{and} \qquad a = \frac{\varphi}{\sqrt{2}} r\cos\delta \quad (\mathrm{A}_5\ \mathrm{T}_e\text{–}\mathrm{S}_1),$$

where $\varphi = \frac{1+\sqrt{5}}{2}$ is the golden ratio, and we expect these to be distinguishable with a superbeam for most of the parameter space [1].

4. Summary

The next generation of neutrino oscillation experiments, with their focus on precision measurements of the underlying parameters, will allow certain classes of models with discrete flavour symmetries to be thoroughly tested. In Ref. [1], the role of a long-baseline superbeam experiment (modelled after LBNO or LBNE) has been shown to be able to exclude these correlations for a large fraction of parameter space. In this contribution, we have highlighted the potential for experimental exclusion of these models at a circa 50 km reactor experiment based on the JUNO facility. By testing the solar predictions to high accuracy, such a facility will be able to independently distinguish between almost all models under consideration. The complementarity between reactor and long-baseline experiments will provide a stringent test of the idea that residual symmetries are responsible for the structure of the PMNS matrix.

References

[1] P. Ballett, S. F. King, C. Luhn, S. Pascoli and M. A. Schmidt, Phys. Rev. D **89** (2014) 016016.
[2] S. F. King, Phys. Lett. B **659** (2008) 244.
[3] D. Hernandez and A. Y. Smirnov, Phys. Rev. D **86** (2012) 053014.
[4] Y.-F. Li, J. Cao, Y. Wang and L. Zhan, Phys. Rev. D **88** (2013) 1, 013008.
[5] S.-B. Kim, talk given at SISAC: from Particle to Universe, SKKU. December 2-4, 2013.

Latest results of NEMO-3: New limit on the $0\nu\beta\beta$ decay half-life for ^{100}Mo

S Blot[1], J Mott[2] and C Vilela[2] (for the NEMO collaboration)

[1]University of Manchester, Oxford Rd, Manchester M13 9PL
[2]University College London, Gower St, London WC1E 6BT

E-mail: blotsd@hep.manchester.ac.uk

Abstract. The NEMO-3 experiment is an experiment dedicated to the search for neutrinoless double β decay ($0\nu\beta\beta$). Detection of $0\nu\beta\beta$ decay would be direct evidence that neutrinos are Majorana particles and lepton number is violated. In NEMO-3, the separation of source isotopes from the active detector region allows for reconstruction of the full event topology, which aids in background suppression and the investigation of $0\nu\beta\beta$ decay underlying mechanisms. NEMO-3 has investigated a total of seven isotopes, ^{100}Mo, ^{82}Se, ^{116}Cd, ^{150}Nd, ^{48}Ca, ^{96}Zr, and ^{130}Te, and has produced the world's best $2\nu\beta\beta$ decay half-life measurements for all of them. No indication of $0\nu\beta\beta$ decay has yet been observed. Presented here is the newly published limit of $T_{1/2}^{0\nu\beta\beta} > 1.1 \times 10^{24}$ y (90% C.L.) for 37 kg·y exposure of ^{100}Mo.

1. Introduction

The NEMO-3 experiment searches for neutrino-less double beta decay ($0\nu\beta\beta$). Observation of $0\nu\beta\beta$ would be direct evidence that neutrinos are Majorana particles and lepton number conservation is violated. Therefore, $0\nu\beta\beta$ decay is a process which is beyond the current Standard Model of particle physics [1]. The experimental signature of this decay is the emission of two electrons whose total energy is equal to the kinetic energy available in the decay ($Q_{\beta\beta}$). The largest background to this process is the $2\nu\beta\beta$ decay which is allowed by the Standard. If the $0\nu\beta\beta$ decay process is mediated by a massive Majorana neutrino, then the rate of the decay is related to the effective neutrino mass, $\langle m_\nu \rangle$, by the equation

$$T_{1/2}^{0\nu\beta\beta} = G^{0\nu}|M^{0\nu}|^2|\langle m_\nu \rangle|^2, \quad (1)$$

where $G^{0\nu}$ and $M^{0\nu}$ are the theoretically calculated phase space and nuclear matrix elements, respectively.

2. The NEMO-3 experiment

The NEMO-3 experiment was located in the Modane Underground Laboratory (LSM), and operated from February 2003 to January 2011. It is unique among $0\nu\beta\beta$ decay experiments due to the separation of the $2\nu\beta\beta$ decay isotopes from the active detector system as shown in Figure 1. It is cylindrical in shape, with the $2\nu\beta\beta$ decay source foils located at a fixed radius. The foils are surrounded by two wire tracking chambers, which provide 3D position measurements of charged particles passing through the detector volume. Surrounding the tracker on all sides is the calorimeter system, composed of plastic scintillator blocks coupled to low-radioactivity 3" and 5" photomultiplier tubes (PMT), which provides both timing and energy measurements. More information about the detector can be found in [2]. The separated tracker-calorimeter system allows for direct measurements of internal and external background rates and final state kinematic variables. Seven different isotopes were investigated *simultaneously* for $0\nu\beta\beta$ and $2\nu\beta\beta$ decay with the NEMO-3 detector. The most massive isotope investigated was ^{100}Mo with 6.9 kg.

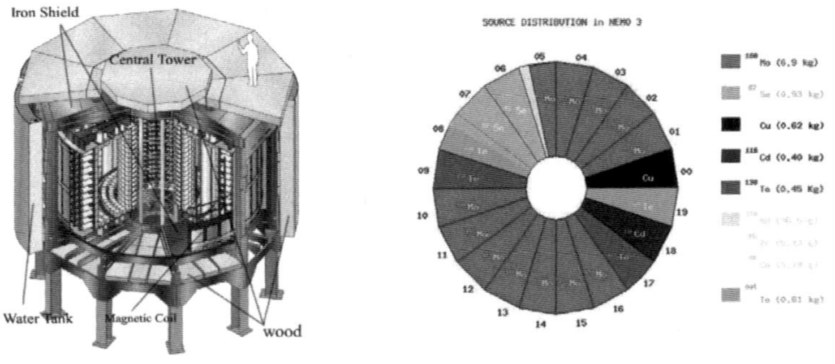

Figure 1. The NEMO-3 detector is shown on the left, and top down view of the layout of the seven different $2\nu\beta\beta$ decay isotopes investigated by NEMO-3 is on the right. ^{100}Mo (red) was the most massive isotope at 6.9 kg.

3. Measuring backgrounds with NEMO-3

The source foils contain small amounts of radioactive isotopes which were detected with a high-purity Germanium (HPGe) detector before installation in the detector. These isotopes may decay via the emission of one electron or positron, and some gamma rays. These decays can result in a two electron final state through many processes including pair production, Compton and Møller scattering, and emission of a β particle together with a conversion electron. Using a number of control regions defined by different final state topologies, the activity of background isotopes, both internal and external to the source foil, can be directly measured with the NEMO-3 detector, compared with the HPGe measurements when possible, and used to estimate the number of expected events in the $2\nu\beta\beta$ and $0\nu\beta\beta$ signal regions. The most significant background to the ^{100}Mo $0\nu\beta\beta$ decay search is the irreducible $2\nu\beta\beta$ decay spectrum. The half-life for this process as measured by NEMO-3 is $T_{1/2}^{2\nu\beta\beta} = (6.93 \pm\ 0.04\ (\text{stat}))\times 10^{18}$ years [3].

4. Results of the $0\nu\beta\beta$ search for ^{100}Mo

The full NEMO-3 data set is used to search for the $0\nu\beta\beta$ decay of ^{100}Mo to the ground state of ^{100}Ru. Events are selected which have two final state electrons with a summed energy in the region 2.8-3.2 MeV. More details on the selection criteria applied to the data set can be found in Reference [3]. The signal efficiency in the region of interest is 4.7%. The number of expected background events are shown in Table 1.

No events above background are observed. Therefore, using an exposure of 34.7 kg·y of ^{100}Mo, a 90% CL limit is set at $T_{1/2}^{0\nu\beta\beta} > 1.1 \times 10^{24}$ years using the summed electron energy spectrum shown in Figure 2. This limit is obtained using the CLs method, where both statistical and systematic errors have been taken into account. Assuming the exchange of a massive Majorana neutrino, this corresponds to a limit on the effective neutrino mass of $\langle m_\nu \rangle < 0.3$-$0.9$ eV when the $M^{0\nu}$ and $G^{0\nu}$ are taken into account [3].

Background	Expected events
Externals	<0.2
Radon	5.2 ± 0.5
^{214}Bi internal	1.0 ± 0.1
^{208}Tl internal	3.3 ± 0.3
^{100}Mo $2\nu\beta\beta$	8.45 ± 0.05
Total background	18.0 ± 0.6
Data	15

Table 1. Number of expected background and data events in the search window of 2.8-3.2 MeV.

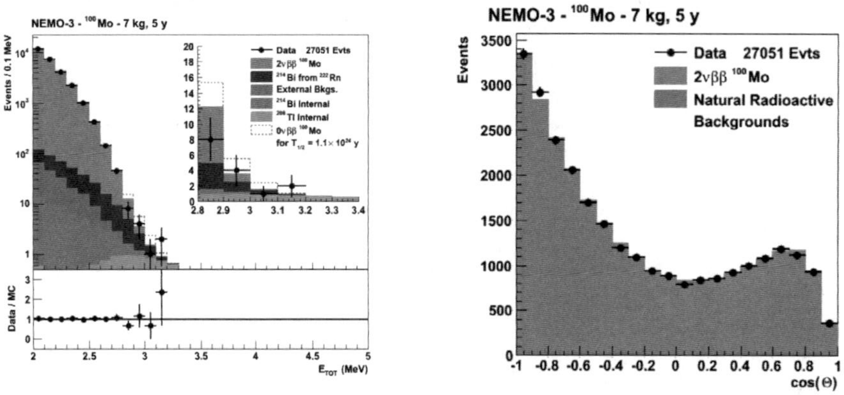

Figure 2. The summed energy (left) and the cosine of the angle between the tracks (right) of the two electrons for the final selection of ^{100}Mo 2e events. The red dashed line indicates the limit set by NEMO-3.

5. Conclusion

The NEMO-3 detector has been used to search for $0\nu\beta\beta$ of seven different isotopes simultaneously. No signal has been observed, and the best NEMO-3 limit using 34.7 kg·y of ^{100}Mo has been set at $T_{1/2}^{0\nu\beta\beta} > 1.1 \times 10^{24}$ years (90% C.L.). Updated half-life values for $2\nu\beta\beta$ decay and limits for $0\nu\beta\beta$ decay as well as other rare and exotic decays, such as $2\nu\beta\beta$ to excited states of daughter isotopes, using the full data set from NEMO-3 for ^{82}Se, ^{116}Cd, ^{150}Nd, ^{48}Ca, and ^{96}Zr are currently under way and nearing completion. Future results hope to improve upon the current published half-life measurements and limits for these processes by utilizing many final state observables from both signal and background channels.

References

[1] Rodejohann W 2011 Neutrinoless Double Beta Decay and Particle Physics *Int. J. Mod. Phys.* **E20** 1833-1930
[2] Arnold R *et al* 2005 Technical design and performance of the NEMO-3 detector *Nucl. Instrum. Meth.* A **536** 79-122
[3] Arnold R *et al* 2014 Search for neutrinoless double-beta decay of Mo100 with the NEMO-3 detector *Preprint* arXiv:1311.5695 (accepted for publication in *Phys. Rev.* D)

Confidence in the Mass Hierarchy Determination from Reactor Neutrino Experiments

E Ciuffoli[1], **J Evslin**[2] **and X M Zhang**[1]

[1] Institute of High Energy Physics (IHEP), CAS, Beijing 100049, China
[2] Institute of Modern Physics, CAS, NanChangLu 509, Lanzhou 730000, China

E-mail: `ciuffoli@ihep.ac.cn`

Abstract. In the next decade, a number of experiments will attempt to determine the neutrino mass hierarchy. We will discuss some of the challenges that these experiments will face, focusing on the reactor neutrino experiments at intermediate baselines. Here, for example, even a small nonlinear energy response can severely affect the sensitivity to the hierarchy. Using MC simulations, we will show the probability of successfully determining the hierarchy for some realistic locations of the detector. We will also show that, since the two hierarchies are non-nested hypothesis, the statistic $\Delta\chi^2$ does not follow a one-degree-of-freedom χ^2 distribution and hence the sensitivity to the hierarchy cannot be estimated by taking the square root of the expected $\Delta\chi^2$; we will present the correct formula for the sensitivity.

1. Introduction

The recent measurement of θ_{13} revealed that it is significantly larger than the expected. One of the consequences of this discovery is that now it is practical to determine the neutrino mass hierarchy using reactor neutrino experiments at intermediate baselines. We will present some of the challenges that this kind of experiment will face, discussing possible ways to increase the sensitivity to the hierarchy. More in detail, we will present our results on the interference effect, which happens when the detector receives neutrinos from two or more nuclear reactors, on the non-linear energy response of the detector and we will discuss about the statistical distribution of $\Delta\chi^2$ in the hierarchy determination.

2. Interference

Usually a detector receives neutrino from different nuclear reactors; the different baselines mean that the oscillation probabilities are out of phase, triggering an interference effect [1]. If ΔL is of the order of 1 km the 1-3 oscillations are affected and the sensitivity to the hierarchy can be drastically reduced; if the difference between the baselines is larger (>10 km) the probability of determine the right hierarchy is not so compromised, but it could affect the precise measurement of θ_{12}, which is another of the goals of this kind of experiment.

Using Monte Carlo simulations we studied the effect of the interference for different values of ΔL and we compared the probability of determine the right hierarchy for several realistic locations of the detector [2, 3]. For the JUNO detector, we proposed a new detector site, DongKeng, between the TaiShan and YangJiang reactor complexes. Our Monte Carlo study revealed that it leads to a much better sensitivity to the hierarchy than the old Daya Bay site.

3. Statistical distribution of $\Delta\chi^2$

The determination of the hierarchy is basically the determination of the sign of $\Delta m^2_{31} = m^2_3 - m^2_1$: if it is positive the hierarchy is called normal, otherwise it is inverted.

Usually the Wilks' theorem ensures that the $\Delta\chi^2$ follows a statistical one-degree-of-freedom χ^2 distribution; however in the case of the hierarchy the two hypothesis are non-nested, *i.e.* one is not a subset of the other. In this case the Wilks' theorem cannot be applied, as pointed out first by [4], and the $\Delta\chi^2$ follow a Gaussian distribution with

$$\mu = \overline{\Delta\chi^2} \qquad \sigma = 2\sqrt{\overline{\Delta\chi^2}} \tag{1}$$

where $\overline{\Delta\chi^2}$ is the expected value of $\Delta\chi^2$. We extended this result [5], showing that it is true also if pull parameters are used, and we tested it with Monte Carlo simulations.

3.1. Definitions of Sensitivity to the Mass Hierarchy

In the case of a null hypothesis nested in a general hypothesis, Wilks' theorem yields a frequentist measure of the confidence in the null hypothesis: this number can be obtained simply taking the square root of $\Delta\chi^2$. The neutrino mass hierarchies are disjoint, not nested hypotheses and so Wilks' theorem does not apply. We are instead interested in the sensitivity to the hierarchy, which is the Bayesian probability that the hierarchy predicted by a binary classification test is the true hierarchy. We present simple formulas for the mean and median sensitivities of a χ^2 test to the hierarchy. Although the sensitivity is a probability, we will use the error function to express them as a number of σ's

- The **mean sensitivity** answers to the question: "**which is the probability that a certain experiment will find the correct hierarchy?**". The number s_c of σ's in this case reads

$$s_c = \sqrt{\overline{\Delta\chi^2}/2} \tag{2}$$

- However, it could be more interesting to consider the sensitivity that can be obtain from a "typical" experiment. The **median sensitivity** answers to the question: "**which is the probability that the median experiment will give us the correct hierarchy?**" (we define the median experiment as the one in which $\Delta\chi^2 = \overline{\Delta\chi^2}$). Assuming symmetric Bayesian priors the number s_v of σ's is

$$s_v = \sqrt{2}\mathrm{erf}^{-1}\left(\frac{1 - e^{-\overline{\Delta\chi^2}/2}}{1 + e^{-\overline{\Delta\chi^2}/2}}\right) \tag{3}$$

In Fig. 1 you can find these two quantities plotted for different values of $\overline{\Delta\chi^2}$.

4. Non-Linear Energy Response

Since the positions of the peaks must be determined with great precision, even a small non-linear energy response of the detector can significantly affect the precision of the experiment [6]. We consider three different models of non-linearity [7]: one where ΔE is proportional to E^2, one in which it decreases exponentially and one in which the non-linearity is tuned to mimic the behavior of the inverted hierarchy (we call it the"worst-case" model). We calculated the expected $\Delta\chi^2$ using these models: the results can be found in Fig. 2. We also calculated the same quantity for some realistic locations of the JUNO and RENO 50 detectors, using the worst-case model where the energy shift ΔE was rescaled by a factor 0 (*i.e.* no non-linearity), 1/3, 2/3 and 1, respectively: the results can be found in Tab. 1. It is clear that a second detector can significantly reduce the systematic error due to the non-linear response. Moreover, with two detectors it is possible to measure with good precision the leptonic CP violation with the addiction of a cyclotron complex [8].

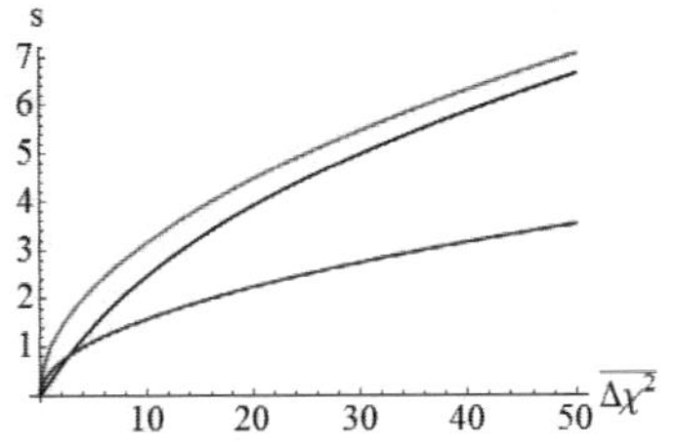

Figure 1. s_c (blue curve) and s_v (black curve). For comparison, also $\sqrt{\Delta\chi^2}$ is shown (red curve).

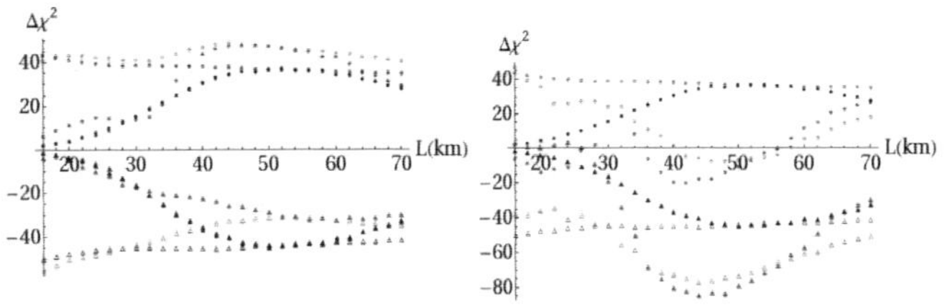

Figure 2. Black: No non-linearity, Blue: quadratic model, Red Panel 1: exponential model, Red Panel 2 :worst-case model. Empty circles (triangle): 1 20ktons detector and normal (inverted) hierarchy; Filled circles (triangles): 2 10ktons detector, one at 55km and normal (inverted) hierarchy. 6 years running.

Table 1. Expected $\Delta\chi^2$ for specific locations of the detectors

Site	NH/IH:0	NH/IH:1/3	NH/IH:2/3	NH/IH:1
DongKeng (JUNO)	14.1/-17.0	8.2/-21.5	2.2/-24.6	-2.4/26.3
+LuGuJing	13.2/-16.2	7.8/-21.4	3.1/-25.6	-2.3/-27.6
+ZiLuoShan	13.5/-16.1	13.9/-15.3	13.7/-15.8	14.2/-14.2
GuemSeong (RENO50)	6.2/-7.7	3.3/-10.0	0.2/-12.1	-2.5/-13.0
+Jangamsan	5.6/-6.6	5.3/-7.0	3.8/-9.4	4.7/-9.8
Munmyeong (RENO50)	11.8/-13.6	6.4/-17.5	1.6/-21.3	-2.9/-23.2
+Buncheon-ri	11.5/-13.6	9.4/-16.4	7.8/-17.3	3.2/-20.2
Munsusan (RENO50)	9.4/-11.7	5.9/-14.4	2.3/-16.8	-1.9/-18.3
+Buncheon-ri	10.3/-12.0	8.6/-14.6	6.9/-15.3	4.6/-16.6

Acknowledgments

EC is supported by the Chinese Academy of Sciences Fellowship for Young International Scientists grant number 2013Y1JB0001 and NSFC grant 11350110500. XZ is supported in part by NSFC grants 11121092, 11033005 and 11375202

References

[1] E Ciuffoli, J Evslin and X M Zhang, JHEP **1303** (2013) 016.
[2] E Ciuffoli, J Evslin and X M Zhang, JHEP **1212** (2012) 004.
[3] E Ciuffoli, J Evslin and X M Zhang, Phys.Rev.**D**88 (2013) 033017.
[4] X Qian, A Tan, W Wang, J J Ling, R D McKeown and C Zhang, Phys.Rev.**D**86 (2012) 113011.
[5] E Ciuffoli, J Evslin and X M Zhang, arXiv:1305.5150 [hep-ph].
[6] E Ciuffoli, J Evslin, Z Wang, C Yang, X M Zhang and W Zhong, arXiv:1211.6818 [hep-ph].
[7] E Ciuffoli, J Evslin, Z Wang, C Yang, X M Zhang and W Zhong, arXiv:1308.0591 [hep-ph].
[8] E Ciuffoli, J Evslin and X M Zhang, arXiv:1401.3977 [hep-ph].

Sensitivity to the neutrino oscillation parameters in the Hyper-Kamiokande experiment

Linda Cremonesi

School of Physics and Astronomy, Queen Mary University of London, UK

E-mail: l.cremonesi@qmul.ac.uk

Abstract. The Hyper-Kamiokande (Hyper-K) [1, 2] detector is a next generation underground water Cherenkov detector that will serve as the far detector of a long-baseline neutrino experiment (T2HK) located in Japan. T2HK is a natural extension of the very successful T2K experiment.

The upgraded facilities at J-PARC will deliver an off-axis narrow band ($\approx$0.6 GeV) (anti)neutrino beam (750 kW-1 MW) directed to Hyper-K that will be used to measure the appearance and disappearance parameters with unprecedented precision as well as potentially discover CP violation in the lepton sector.

If the mass hierarchy is known, Hyper-K is expected to determine the CP phase to better than 19 degrees for all possible values of δ_{CP} and CP violation can be determined at 3σ for 76% of the delta parameter space (considering 5 years exposure to neutrino beam produced by the 1.66 MW J-PARC proton synchrotron) [2].

1. Neutrino oscillations and CP violation

The neutrino mixing matrix (PMNS - Pontecorvo Maki Nakagawa Sakata) translates neutrino mass eigenstates into flavour eigenstates $(\nu_e, \nu_\mu, \nu_\tau)^T = U(\nu_1, \nu_2, \nu_3)^T$.

In the three flavour framework it is parametrised using 3 mixing angles (θ_{12}, θ_{23}, θ_{13}), one Dirac CP phase δ and two Majorana phases (α_1, α_2).

Neutrino oscillation frequencies are determined by mass parameters $\Delta m^2_{21} = m^2_2 - m^2_1$ and $\Delta m^2_{32} = m^2_3 - m^2_2$, where m_j refers to the mass eigenvalues.

Since oscillation probabilities do not depend on the Majorana phases, the number of relevant parameters is 6, of which: $(\Delta m^2_{21} = (7.50 \pm 0.20)10^{-5}eV, \sin^2 2\theta_{12} = 0.857 \pm 0.024$ [3]) have been measured by solar neutrino experiments; $(|\Delta m^2_{32}| = 2.32^{+0.12}_{-0.08}10^{-3}eV, \sin^2 2\theta_{23} > 0.95$ [4]) have been measured by atmospheric neutrino experiments; $\sin^2 2\theta_{13} = 0.095 \pm 0.010$ [5] has been constrained by reactor experiments.

It is then crucial for neutrino physics to measure the CP violation δ phase.

2. Hyper-Kamiokande

The Hyper-Kamiokande (Hyper-K) detector is a next generation underground water Cherenkov detector [1] that aims to study atmospheric and solar neutrinos as well as neutrinos from other astrophysical origins, oscillated neutrinos from an upgraded J-PARC beam, and to observe proton decay.

	Signal		Background					Total
	$\nu_\mu \to \nu_e$	$\bar{\nu}_\mu \to \bar{\nu}_e$	ν_μ CC	$\bar{\nu}_\mu$ CC	ν_e CC	$\bar{\nu}_e$ CC	NC	
ν mode	3016	28	11	0	503	20	172	3750
$\bar{\nu}$ mode	396	2110	4	5	222	396	265	3397

Table 1: Expected events after the ν_e ($\bar{\nu}_e$) appearance selection during ν ($\bar{\nu}$) running mode.

In the baseline design, Hyper-K consists of 2 separate caverns each having an egg-shape cross section 48 meters wide, 54 meters tall, and 250 meters long, resulting in a total (fiducial) mass of 0.99 (0.56) million metric tons (about 20 (25) times larger than that of Super-K).

The two water tanks are then divided into 10 compartments (5 each), each been optically separated and consisting of an inner detector (ID) and an outer detector (OD). The ID is viewed by an inward-facing array of 20-inch diameter PMTs. The entire array consists of 99,000 Hamamatsu R3600 photomultiplier tubes (PMTs), uniformly surrounding the region and giving a photocathode coverage of 20%.

The OD region is equipped with 25,000 8-inch diameter PMTs.

The detector simulation has been studied using WCSim [6], a flexible Geant4-based tool developed by Duke University to study a water Cherenkov detector with top and side photo-multiplier tubes. The egg-shape geometry of Hyper-K has been implemented in WCSim.

3. Tokai to Hyper-Kamiokande

Hyper-K will serve as the far detector of a long-baseline neutrino experiment (T2HK) located in Japan.

Following the lines of T2K, the facilities at J-PARC will deliver an off-axis narrow band ($\approx$0.6 GeV) (anti-)neutrino beam (750kW-1MW) directed to Hyper-K that will be used to measure the appearance and disappearance spectra.

The near detector complex will include the INGRID and ND280, on- and off-axis near detectors respectively, located 280 m away from the beam target. They are the current T2K near detectors, but will be upgraded for T2HK. A new near detector will possibly be built at 2 km from the target. The near detectors will help constrain the neutrino flux and systematic errors.

Table 1 shows the expected number of events after the $\nu_e/\bar{\nu}_e$ appearance selection. In ν running mode the main background is composed of beam ν_e CC and all NC interactions. In $\bar{\nu}$ running mode, the wrong sign appearance is also non-negligible.

The reconstructed neutrino energy distributions of ν_e events for several values of δ_{CP} are shown in the top plots of Figure 1. The bottom plots show the difference of reconstructed energy spectrum from $\delta_{CP} = 0$, for the cases $\delta_{CP} = \pi/2, \pi, 3\pi/2$. By using not only the total number of events but also the reconstructed energy distribution, the sensitivity to δ_{CP} can be improved, and one can discriminate all the values of δ_{CP}, including the difference between $\delta_{CP} = 0$ and π.

4. CP violation discovery sensitivity

Sensitivity studies have been performed taking into account an integrated beam power of 7.5 MW $\times 10^7$ sec (corresponding to 1.56×10^{22} protons on target with 30 GeV J-PARC beam) and a 1:3 ratio of neutrino to anti-neutrino running.

Figure 2(left) shows the expected significance to exclude $\sin\delta_{CP} = 0$ (CP conserved case) with different systematic uncertainty hypotheses. The significance is calculated as $\sigma = \sqrt{\Delta\chi^2}$, where $\Delta\chi^2$ is the difference of the trial value of δ_{CP} and $\delta_{CP} = 0$ or 180 (the smaller value of the difference is taken).

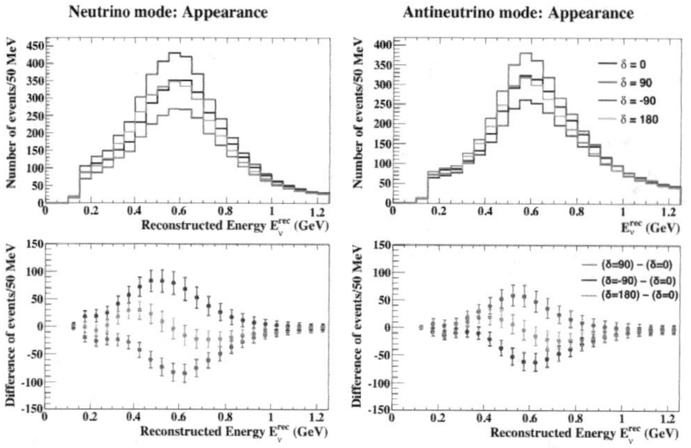

Figure 1: Top: Reconstructed neutrino energy distribution for several values of δ_{CP}. Bottom: Difference of the reconstructed neutrino energy distribution from the case with $\delta_{CP} = 0$. The error bars represent the statistical uncertainties of each bin.

Figure 2 (centre) shows the fraction of δ_{CP} for which $\sin\delta_{CP} = 0$ can be excluded at 3σ for different values of $\sin^2 2\theta_{13}$: if $\sin^2(2\theta_{13}) > 0.03$, CP violation in the lepton sector can be observed with 3σ for 76% of the delta parameter space.

Figure 2 (right) shows the 1σ uncertainty for δ_{CP} as a function of the integrated beam power: considering full Hyper-K statistics (i.e. 5 years exposure to neutrino beam produced by the 1.66 MW J-PARC proton synchrotron) the δ_{CP} phase could be determined to better than 19 degrees for all possible values of δ.

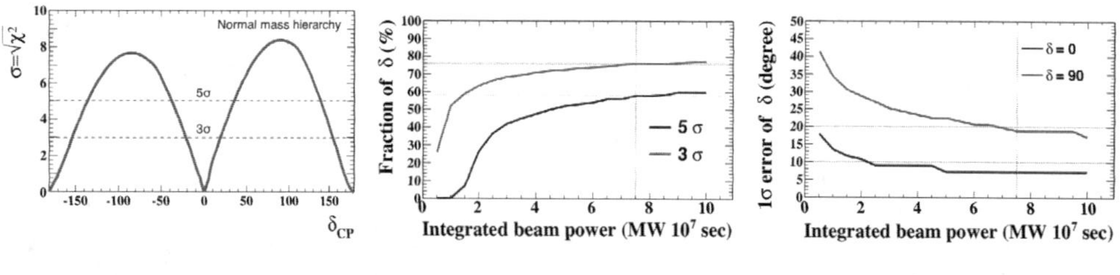

Figure 2: Expected significance to exclude $\sin\delta_{CP} = 0$ (left). Fraction of δ_{CP} for which $\sin\delta_{CP} = 0$ can be excluded at 3σ for different values of $\sin^2 2\theta_{13}$ (centre). 1σ uncertainty for δ_{CP} as a function of the integrated beam power (right).

References

[1] Abe K *et al* 2011 Letter of Intent: The Hyper-Kamiokande Experiment—Detector Design and Physics Potential *arXiv preprint* hep-ex 1109.3262
[2] Hyper-Kamiokande Working Group, Abe K *et al* 2014 A Long Baseline Neutrino Oscillation Experiment Using J-PARC Neutrino Beam and Hyper-Kamiokande *arXiv preprint* physics.ins-det 1412.4673
[3] Abe K *et al* 2011 Search for differences in oscillation parameters for atmospheric neutrinos and antineutrinos at Super-Kamiokande *Phys. Rev. Letters* **107** 241801
[4] Adamson P *at al* 2011 Measurement of the neutrino mass splitting and flavor mixing by MINOS *Phys. Rev. Letters* **106** 181801
[5] Beringer J and Particle Data Group *et al* 2012 *Phys. Rev. D* **86** 010001
[6] GitHub repository for WCSim: https://github.com/WCSim/WCSim

Journal of Physics: Conference Series **598** (2015) 012018 doi:10.1088/1742-6596/598/1/012018

Muon neutrino disappearance at T2K

T Dealtry[1,2] for the T2K collaboration

[1] STFC, Rutherford Appleton Laboratory, Harwell, Oxford, UK

[2] Department of Physics, University of Oxford, Oxford, UK

E-mail: thomas.dealtry@physics.ox.ac.uk

Abstract. The T2K experiment has observed the disappearance of muon neutrinos from the J-PARC muon-neutrino beam. Using data corresponding to 3.01×10^{20} protons on target, 58 muon neutrino event candidates were observed; 205 ± 17 (syst.) events are expected under the no-oscillation hypothesis. A fit to the observed rate and energy spectrum in a neutrino oscillation framework with three active neutrino flavours in constant density matter, assuming the normal mass hierarchy, yields a best fit mixing angle $\sin^2\theta_{23} = 0.514 \pm 0.082$ and best fit mass squared splitting $|\Delta m^2_{32}| = 2.44^{+0.17}_{-0.15} \times 10^{-3}\,\mathrm{eV}^2/\mathrm{c}^4$, which corresponds to the maximal disappearance probability.

T2K is an off-axis long-baseline neutrino experiment using the J-PARC muon-neutrino beam to study neutrino oscillation parameters. In the limit $\Delta m^2_{21} \ll |\Delta m^2_{32}|$, the probability of muon neutrino disappearance is

$$P(\nu_\mu \to \nu_\mu) \simeq 1 - 4\cos^2\theta_{13}\sin^2\theta_{23}[1 - \cos^2\theta_{13}\sin^2\theta_{23}]\sin^2(1.267\Delta m^2_{32}L/E_\nu), \quad (1)$$

where L(km) is the neutrino propagation distance, E_ν(GeV) is the neutrino energy, and $\Delta m^2_{32}(\mathrm{eV}^2/\mathrm{c}^4)$ is the neutrino mass-squared splitting, $\Delta m^2_{32} = m^2_3 - m^2_2$. From equation 1, it can be shown that maximal disappearance occurs at $\sin^2\theta_{23} \simeq 1/(2 - 2\sin^2\theta_{13})$, and the disappearance probability is symmetric in $\sin^2\theta_{23}$ about this point. Using recent measurements of θ_{13} ($\sin^2(2\theta_{13}) = 0.098$ [1]), maximal disappearance occurs at $\sin^2\theta_{23} \approx 0.513$. Since maximal disappearance occurs at a point which is not the same as maximal 23-sector mixing ($\sin^2\theta_{23} = 0.500$) it is also necessary to fit for $\sin^2\theta_{23}$, rather than the historically used $\sin^2(2\theta_{23})$, since the oscillation probability depends on the choice of octant ($\sin^2\theta_{23} < 0.5$ or $\sin^2\theta_{23} > 0.5$).

The T2K experiment consists of a muon-neutrino beam, a suite of near detectors, and a far detector, Super-Kamiokande (SK). The muon-neutrino beam is generated using a 30 GeV proton beam produced at J-PARC. Three magnetic horns are used to focus charged pions and kaons produced in interactions of the beam with a graphite target, which then decay in a 96-m-long decay volume. SK is a 22.5 kt fiducial volume water Cherenkov detector located 2.5° off-axis and 295 km from the graphite target. The off-axis angle means SK sees a narrow-band neutrino beam, with energy $E_\nu \approx 600\,\mathrm{MeV}$, corresponding to the first oscillation maximum at $E_\nu = 1.267\Delta m^2_{32}L(2/\pi)$. This enhances the sensitivity to the oscillation parameters by increasing the number of neutrinos that oscillate, and decreasing backgrounds from the high-energy tail. The unoscillated neutrino beam is sampled by ND280, a near detector 2.5° off-axis and 280 m from the graphite target, in order to constrain the flux and cross section models. Here we report the results of the ν_μ disappearance search with T2K Runs 1-3, consisting of 3.01×10^{20} protons on target [2].

Fits to external datasets are used to tune both the initial flux [3] and cross section [4] models, and to set the magnitude of the uncertainties. Simulations of proton beam interactions with the graphite target predict the neutrino fluxes at ND280 and SK, along with a covariance between neutrino energies, neutrino flavours, and detectors. Flux uncertainties are dominated by hadron production uncertainties, and are 10-20% in the flux peak region. There are two classes of cross section model parameters, those common between ND280 and SK, and those independent of ND280. This distinction is necessary because interactions at ND280 are predominately on carbon targets, whereas interactions at SK are predominately on oxygen targets. The ND280-common cross section systematics include axial mass parameters for charged current quasi-elastic (CCQE) and resonant pion production, and five normalisation parameters. The independent parameters include parameters controlling the CCQE nuclear model, pion production parameters, and normalisations for samples that are not included in the ND280 fit (including neutral current (NC), antineutrinos and electron neutrinos). A further set of model uncertainties from pion interactions within the nucleus (final-state interactions) and within the detector (secondary interactions) are assigned by varying model parameters to span external datasets.

An inclusive selection of ν_μ CC events is performed at ND280 by searching for the highest-momentum negatively-charged track in the second time projection chamber (TPC2), requiring that the energy loss is consistent with a muon, and requiring it to start in the fiducial volume of the first fine grained detector (FGD1). A veto is applied on TPC1 activity, upstream of FGD1, to reject events with vertices upstream of FGD1. The inclusive sample is split into two subsamples based on the presence of a pion: the CCQE sample contains events with exactly one track matched between FGD1 and TPC2 and no delayed activity in FGD1 consistent with a Michel electron; the CCnonQE sample contains all other events. The purity of the CCQE and CCnonQE samples is 69.5% and 70.4% respectively. The absolute muon momentum scale, pion secondary interactions, and background uncertainties are the largest detector systematics.

The ND280 data are fit in 40 bins (5 momentum $\times$ 4 angular $\times$ 2 sample bins). ND280 flux parameters, both common and ND280 cross section parameters, and detector uncertainties are included in the fit, in order to constrain the flux and cross section models for the oscillation fit. The best-fit values of the 23 flux and common cross section parameters, and their covariance is used in the oscillation fit.

Muon neutrino CCQE events are selected at SK by searching for exactly 1 μ-like Cherenkov ring in a fully-contained event with a fiducial volume vertex, with visible energy $> 30\,\text{MeV}$, reconstructed muon momentum $> 200\,\text{MeV}$, and ≤ 1 reconstructed Michel electron. Timing cuts are also applied such that the event is consistent with the neutrino beam. Observed muon kinematics are used with the quasi-elastic formula to determine the reconstructed neutrino energy

$$E_{\text{reco}} = \frac{m_p^2 - (m_n - E_b)^2 - m_\mu^2 + 2(m_n - E_b)E_\mu}{2(m_n - E_b - E_\mu + p_\mu \cos\theta_\mu)}, \tag{2}$$

where p_μ, E_μ, $\cos\theta_\mu$ are the reconstructed muon momentum, energy, and the angle with respect to the beam direction, respectively, m_p, m_n, and m_μ are the masses of the proton, neutron, and muon, respectively, and $E_b = 27\,\text{MeV}$ is the average binding energy of a nucleon in ^{16}O. Under the no-oscillation hypothesis 205 ± 17 events are predicted, and the spectrum, shown in figure 1, is 77.7% pure in $\nu_\mu + \overline{\nu_\mu}$ CCQE events. SK detector systematics are assigned by comparing both cosmic muon and atmospheric neutrino data with Monte Carlo predictions. The largest uncertainties are due to ring counting, and particle identification for neutral current events.

The error on the number of events from the four error categories at SK is shown in table 1 for a typical oscillation point. The constraint provided by the ND280 fit results in a large reduction in the error (21.8% to 4.2%), resulting in the SK detector systematics dominating (10.1%).

A likelihood ratio fit is performed in order to estimate the oscillation parameters. The fit is performed with 73 variable-width reconstructed energy bins and 48 systematic parameters,

Table 1. Effect of 1σ systematic parameter variation on the number of 1-ring μ-like events, computed for oscillations with $\sin^2\theta_{23} = 0.500$ and $|\Delta m^2_{32}| = 2.40 \times 10^{-3}\,\mathrm{eV}^2/\mathrm{c}^4$.

Source of uncertainty (no. of parameters)	$\delta n^{\mathrm{exp}}_{\mathrm{SK}}/n^{\mathrm{exp}}_{\mathrm{SK}}$
ND280-independent cross section (11)	6.3%
Flux & ND280-common cross section (23)	4.2%
(without ND280 fit) (23)	(21.8%)
SK detector systematics (8)	10.1%
Final-state and secondary interactions (6)	3.5%
Total (48)	13.1%

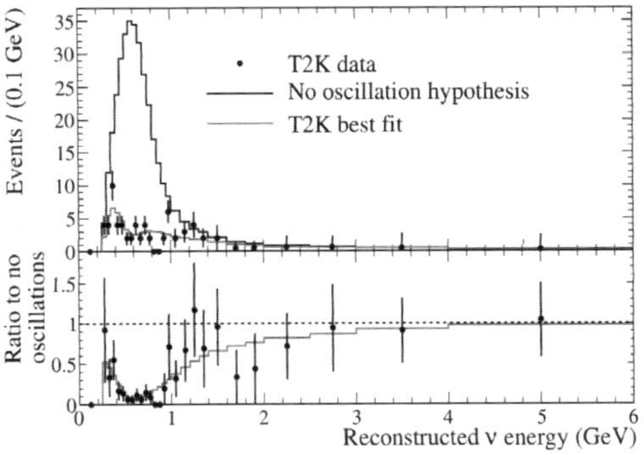

Figure 1. Top: The 58 event 1 μ-like ring SK reconstructed energy spectrum with the prediction for the no-oscillation hypothesis and the best-fit from the oscillation analysis. Bottom: The ratio of both the observed spectrum and best-fit curve to the no-oscillation hypothesis.

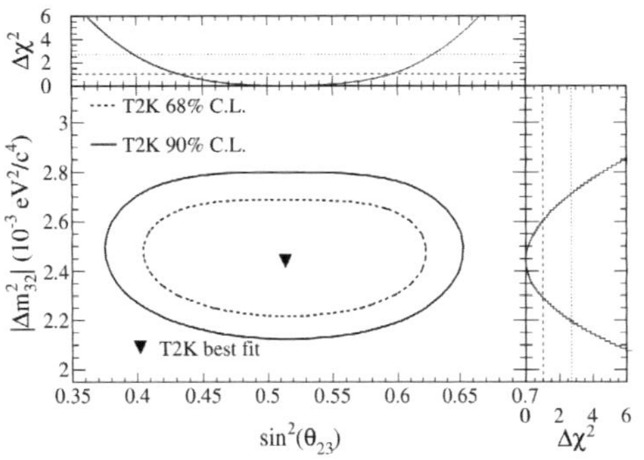

Figure 2. The 68% and 90% C.L. regions for $\sin^2\theta_{23}$ (top), $|\Delta m^2_{32}|$ (right), and $\sin^2\theta_{23}$ versus $|\Delta m^2_{32}|$ (bottom left).

using an oscillation probability generated for three active neutrino flavours in constant density matter of $2.6\,\mathrm{g/cm}^3$. Other oscillation parameters are fixed: $\sin^2(2\theta_{13}) = 0.098$; $\Delta m^2_{21} = 7.5 \times 10^{-5}\,\mathrm{eV}^2/\mathrm{c}^4$; $\sin^2(2\theta_{12}) = 0.857$ [1], $\delta_{CP} = 0$, and the normal mass hierarchy is assumed. The best-fit point is at $\sin^2\theta_{23} = 0.514 \pm 0.082$ and $|\Delta m^2_{32}| = 2.44^{+0.17}_{-0.15} \times 10^{-3}\,\mathrm{eV}^2/\mathrm{c}^4$, and the best-fit spectrum is shown in figure 1. Confidence regions are constructed with the constant-$\Delta\chi^2$ method [1], and the 1-dimensional and 2-dimensional regions are shown in figure 2. This result corresponds to the maximum disappearance probability. In the time since this poster was presented, T2K has published new results using more than twice as much data, and an improved ND280 analysis [5].

References

[1] Beringer J *et al.* (Particle Data Group) 2012 *Phys.Rev.* D **86** 010001
[2] Abe K *et al.* (T2K Collaboration) 2013 *Phys.Rev.Lett.* **111** 211803
[3] Abe K *et al.* (T2K Collaboration) 2013 *Phys.Rev.* D **87** 012001
[4] Hayato Y 2009 *Acta Phys.Polon.* B **40** 2477–2489
[5] Abe K *et al.* (T2K Collaboration) 2014 *Preprint* arXiv:1403.1532 [hep-ex]

Radiopurity requirements for the SuperNEMO experiment and the BiPo detector

Guillaume Eurin[1,2], on behalf of the SuperNEMO collaboration
[1] UCL, London WC1E 6BT, United Kingdom
[2] LAL, Univ Paris-Sud, CNRS/IN2P3, F-91405 Orsay, France
E-mail: eurin@lal.in2p3.fr

Abstract. The main goal of the SuperNEMO collaboration is to try to observe neutrinoless double-β decay. This would prove that the neutrino is a Majorana particle ($\nu = \overline{\nu}$). Today the best lower limits on half-lives of this process are set around 10^{24} - 10^{26} years as obtained by the NEMO-3 experiment [1] (for the 2β isotope ^{100}Mo) and other experiments. SuperNEMO is the next generation experiment based on the NEMO-3 tracker-calorimeter detection principle. The targeted contamination levels for the source foils are lower than can be measured through γ spectroscopy. A more sensitive detector has been constructed to measure low contaminations in ^{208}Tl (around few μBq/kg) and ^{214}Bi (few dozen μBq/kg) in thin materials: the BiPo detector. BiPo-3 has been fully operational at the *Laboratorio Subterráneo de Canfranc (LSC)* since January, 2013. The construction, performance and calibration of the BiPo-3 detector will be covered as well as the radiopurity requirements for SuperNEMO.

1. Introduction

Since neutrino oscillations have been clearly established, neutrinos are now known to be massive but their masses are very small compared to all the other Standard Model fermions. The current best experimental technique for proving the Majorana nature of the neutrino is the search for the neutrinoless double-β decay ($0\nu2\beta$).

After the success of the NEMO-3 detector [2], a new generation experiment has been designed. The successor of NEMO-3, SuperNEMO [3], has started with the construction of the first module. This demonstrator aims to prove the required performance for SuperNEMO. Data taking is expected to start at the begining of 2016.

2. Radiopurity requirements for SuperNEMO

The typical available energy in the $0\nu2\beta$ decay is $\sim$ 3 MeV. ^{214}Bi and ^{208}Tl can be found in natural radioactivity chains. They are the main backgrounds for $0\nu2\beta$ from inside the $\beta\beta$ source foil. During their single-β decay they emit high-energy particles: $Q_\beta(^{214}\text{Bi}) = 3.27$ MeV and $Q_\beta(^{208}\text{Tl}) = 4.99$ MeV. These isotopes can mimic a double-β decay event via several processes following the single-β decay: Moller scattering of the electron, internal conversion and Compton scattering of the associated γ rays.

The target sensitivity with ^{82}Se is $T_{1/2}(\beta\beta0\nu) > 10^{26}$ years for an exposure of 500 kg.year. To reach this value, the limits on the contamination levels and the energy resolution of the optical modules have been improved. Reducing the value of the energy resolution allows a

Table 1. NEMO-3 parameters and requirements of SuperNEMO.

	NEMO-3	SuperNEMO
Mass	7 kg	100 kg
Isotope	^{100}Mo	^{82}Se (^{150}Nd, ^{48}Ca)
Energy resolution	FWHM = 15% at 1 MeV	FWHM = 8% at 1 MeV
Radon contamination	A(^{222}Rn) = 5 mBq/m^3	A(^{222}Rn) $<$ 0.15 mBq/m^3
^{208}Tl and ^{214}Bi contamination	A(^{208}Tl) = 100 μBq/kg A(^{214}Bi) $\sim$ 300 μBq/kg	A(^{208}Tl) $<$ 2 μBq/kg A(^{214}Bi) $<$ 10 μBq/kg

better discrimination between the $0\nu2\beta$ and the $2\nu2\beta$ decay signals. The half-life sensitivity target of SuperNEMO gives rise to the detector requirements listed in Table 1.

3. The BiPo processes in natural radioactivity chains

SuperNEMO requirements for the contaminations in ^{208}Tl and ^{214}Bi cannot be measured, using non-destructive techniques, with the current performance of High Purity Germanium detectors (HPGe). A new detector has therefore been developed after 5 years of R&D: the BiPo detector which takes advantage of the BiPo cascades that exist in natural radioactivity chains.

A BiPo event consists of a β-decay and a delayed α-decay consistent with the characteristic half-life of the Po nucleus. The energies and delay times for ^{212}BiPo and ^{214}BiPo events are:

- ^{212}Bi : $Q_\beta(^{212}\text{Bi})$ = 2.25 MeV, $E_\alpha(^{212}\text{Po})$ = 8.78 MeV and $T_{1/2}(^{212}\text{Po})$ = 299 ns
- ^{214}Bi : $Q_\beta(^{214}\text{Bi})$ = 3.27 MeV, $E_\alpha(^{214}\text{Po})$ = 7.69 MeV and $T_{1/2}(^{214}\text{Po})$ = 164 μs

4. The BiPo-3 detector

Studies with several prototypes lead to the design of the BiPo-3 detector [4]. BiPo-3 is an ultrapure detector with an effective surface area of 3.6 m^2 capable of measuring 1.4 kg of 2β isotope in a thin source foil (40 mg/cm^2). Each one of the two modules is made of 40 paired light lines (two PMTs + light guides + scintillators).

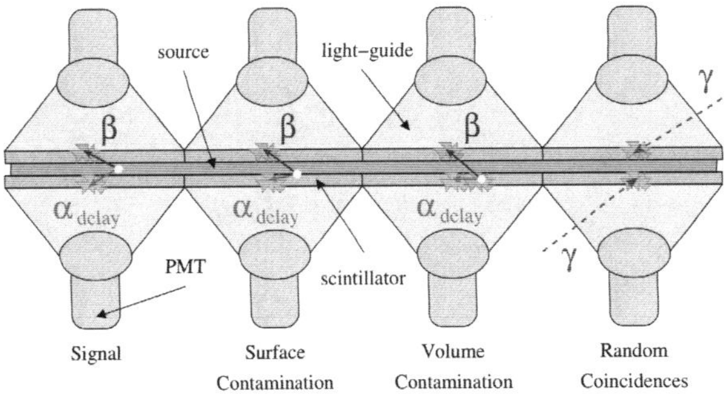

Figure 1. BiPo & background events topologies.

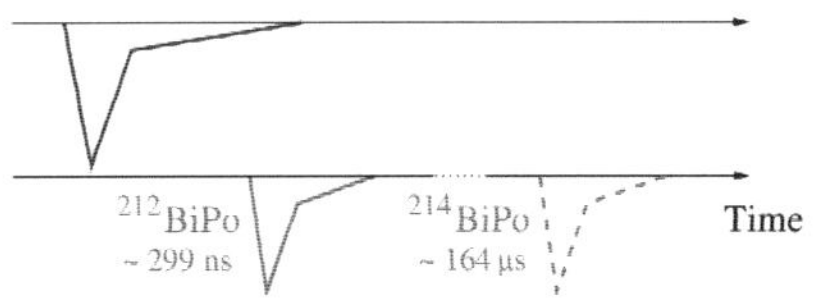

Figure 2. BiPo events as recorded by the acquisition system.

Thanks to the BiPo event's time signature, few backgrounds remain (see figure 1):

- Surface contamination of the scintillators and radon ($T_{1/2}(^{222}\text{Rn})$ = 3.8 d allowing it to diffuse into the detector).
- Volume contamination of the scintillators.
- γ-induced random coincidences from the laboratory or detector components.

The electronics digitizes the PMT pulses (see figure 2) at 1 GS/s and the delay time between events is studied up to 1 ms for the ^{214}BiPo study. The energy deposited by the particles in the scintillator is proportional to the integral of the pulse (charge).

5. Construction and calibration of the BiPo-3 detector

The detector has been installed at the LSC, a deep underground laboratory. It has been assembled in a clean room. The two modules are hosted in a gas tight tank flushed with nitrogen and shielded by iron and lead. The full detector has been running since January 2013.

All the BiPo optical modules have been qualified (PMT gain and energy resolution) before the installation using α (^{241}Am) and β (1 MeV conversion e^- from ^{207}Bi) sources. Regular γ calibrations of the detector are also performed before each measurement. For the BiPo analysis, the energy resolution (FWHM) and the charge-to-energy calibration constant must be determined for each block using statistical tests comparing data and simulations. A time calibration between two opposite optical modules is also performed with crossing electron events, signified by two almost simultaneous energy deposits.

Several corrections were applied to the simulations such as alpha quenching, energy resolution and non-uniformity of the light collection of the scintillator to match the data and the corrected simulations. This was done comparing the electron and α energy spectra.

An aluminium foil has been introduced in the detector as part of the calibration procedure. This foil has a known contamination in both ^{214}Bi and ^{208}Tl (measured in the prototypes and HPGe). An agreement with the expected value of 299 ns on the delay times for the ^{212}BiPo events was observed. For ^{214}BiPo events the value is also compatible with 164 μs. The measured activities agreed with HPGe measurements of the foil.

6. Expected sensitivities

The sensitivities of the BiPo detector for the measurement of the SuperNEMO source foils have been estimated using background measurements (see figures 3 & 4). Dedicated runs were performed to determine the background level of the detector. The required activities have to be reached within $\sim$ 6 months. Some improvements of the analysis are still ongoing, such as a source surface contamination rejection method. However the current sensitivities are at the levels of the requirements of the SuperNEMO demonstrator.

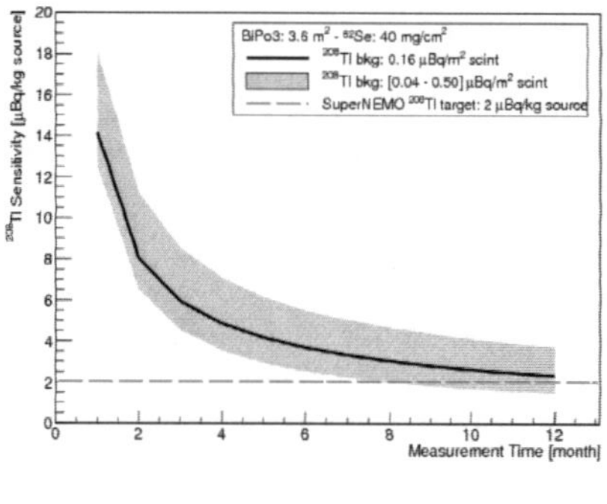

Figure 3. Expected sensitivity in ^{208}Tl.

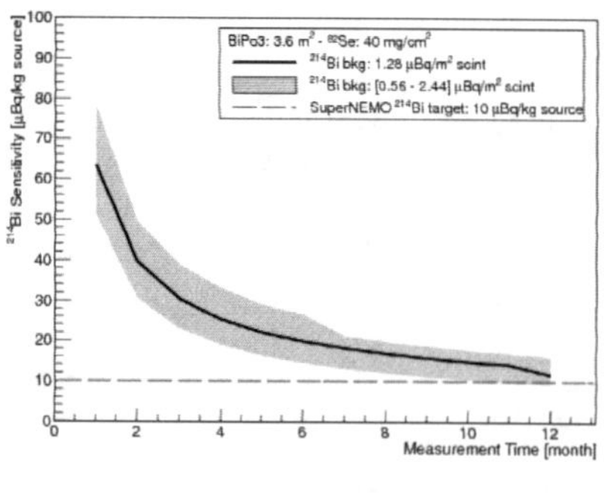

Figure 4. Expected sensitivity in ^{214}Bi.

References

[1] NEMO-3 Collaboration, Accepted to be published in Phys. Rev. D 2014 (arXiv:1311.5695)
[2] NEMO-3 Collaboration, Nuclear Physics A, Volume 781, Issues 1-2, 209-226, 2007 (arXiv:hep-ex/0609058)
[3] SuperNEMO Collaboration, Eur. Phys. J. C70:927-943, 2010 (arXiv:1005.1241)
[4] SuperNEMO Collaboration, NIM A 622, 2010, 120-128 (arXiv:1005.0343)

Construction of the tracker for the SuperNEMO experiment

P Guzowski
University of Manchester, Oxford Road, Manchester, M13 9PL, UK
E-mail: pawel.guzowski@hep.manchester.ac.uk

Abstract. The SuperNEMO experiment will search for neutrinoless double β decay with a sensitivity two orders of magnitude greater than its predecessor NEMO-3, achieved through better energy resolution and unprecedented background levels. A Demonstrator module of the detector is currently under construction, due to begin data taking in 2015. A vital component of the detector is the tracker, a wire chamber made from 2,034 cells operating in Geiger mode. This is important for triggering, background discrimination, and measurement of the kinematic properties of the decay. The tracker production procedure and status is presented.

1. Introduction

The SuperNEMO experiment will search for neutrinoless double β decay, with a half-life sensitivity of around 10^{26} years. It will be made of 20 modules, each one using 5–10 kg of source isotope in a 4×3.7 m^2 foil sheet. A magnetised tracker volume will occupy the space on both sides of the foil. The tracker will be made of 113 rows of 9 tracker cells per side. A calorimeter will surround the tracker, formed of around 500 scintillator blocks coupled to PMTs. The purpose of the calorimeter is to measure the energy and time of the β decay electrons and other background particles. The tracker will be used to discriminate between electrons, positrons, photons and alphas, as well as for measuring event kinematics and topology, and allow for in-situ measurements of background sources. The first of the 20 modules (the 'Demonstrator') is currently being manufactured.

2. The tracker cell

The basic unit of the tracker is the wire cell, shown schematically in Figure 1. This is formed of a 2.7 m long, 40 μm diameter, stainless steel anode wire, surrounded by 8 grounded cathode wires (50 μm diameter), making a 3×3 cell with 2 cm separation. At both ends of the anode wire is a copper cathode end cap. The cell is operated in Geiger mode, with an operating voltage of around 1800 V. Charged particles passing through the cell leave ionisation electrons that drift towards the anode wire. In the high field region close to the wire, further ionisation produces UV light which induces new ionisation further out. This sets up a chain reaction, and a gas of ionised plasma spreads out from the initial track point, parallel to the wire. The time that this plasma reaches the end caps is used to calculate the longitudinal position of the initial track.

More details on the functionally-identical NEMO-3 tracker cell are given in section 2.3.1 of [1].

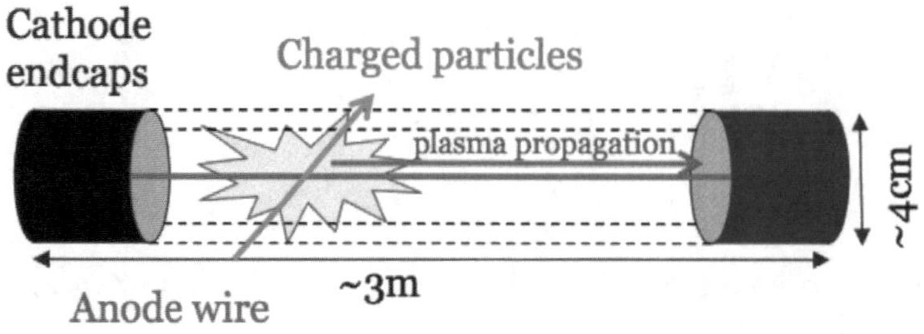

Figure 1. Schematic of a single tracker cell, showing the anode wire (red line), ground cathode wires (dashed lines), and end caps (blue).

3. Tracker production

All components of the cells are cleaned beforehand, in ultrasonic baths, with the wires being spooled through the baths. The components are assembled in a class 1000 clean room, where a specially designed wiring robot has been created for the tracker production task. The robot allows automation of the construction, with minimal human contact that could introduce impurities into the system. It also allows for outsourcing the production to multiple parallel hosts, when the remaining 19 modules of the experiment are built. The robot produces one tracker cell in around 20 minutes.

Eighteen of these cells are assembled into one 'cassette' – a modular unit with two rows of nine cells. This unit allows easy handling for testing, transportation and installation. A total of 113 cassettes will be built for the whole Demonstrator module.

4. Testing

Following construction, the cassettes are tested using cosmic rays, inside an air-tight gas chamber. The gas used is a mixture of 95% helium, 1% argon and 4% ethanol. A measurement of the trigger rate is made while the voltage is scanned, and is used to determine the Geiger plateau, which starts at approximately 1700 V. An example of this measurement is shown in Figure 2. Cells can be rejected at this stage if their cosmic trigger rate is lower than around 40 Hz (indicating an inefficiency), or if the plateau is narrower than 100 V.

Following the plateau measurement, the cells are individually tested at the correct operating voltage (at the centre of the plateau), with 10,000 cosmic triggers taken. This data is used to measure the cell performance: most importantly the cathode efficiency (the fraction of events with a signal at the cathode end caps), the plasma propagation time, and the longitudinal position of each cosmic ray. Examining the plasma propagation time can reveal any dust on the wire which stops the plasma propagation, as there will be a multiple peak structure depending on what side of the dust the cosmic ray was. The longitudinal position should be uniform, so any peaks in this distribution will reveal self-triggering points on the wire (areas where the electric field is strong enough to induce spontaneous ionisation). Examples of these failure-mode distributions are shown in Figure 3. Cells are rejected if they have any of these deficiencies.

5. Conclusion

The construction of the SuperNEMO tracker is well under way. To date, thirteen cassettes have been produced, of which only one has been rejected due to failing the testing procedure. The experiment is on target to begin data taking in late 2015.

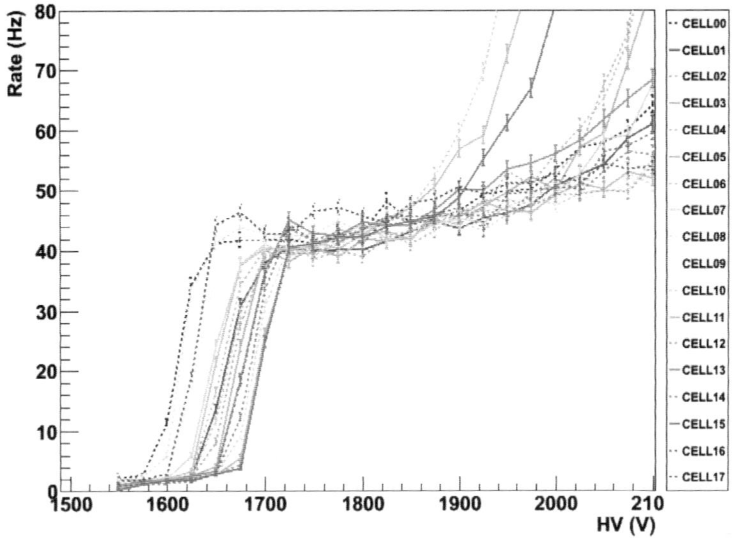

Figure 2. Geiger plateau measurement for 18 cells of a single cassette, showing an approximately constant trigger rate above a threshold voltage. For three of the cells, the plateau is relatively narrow, breaking down at around 1800 V.

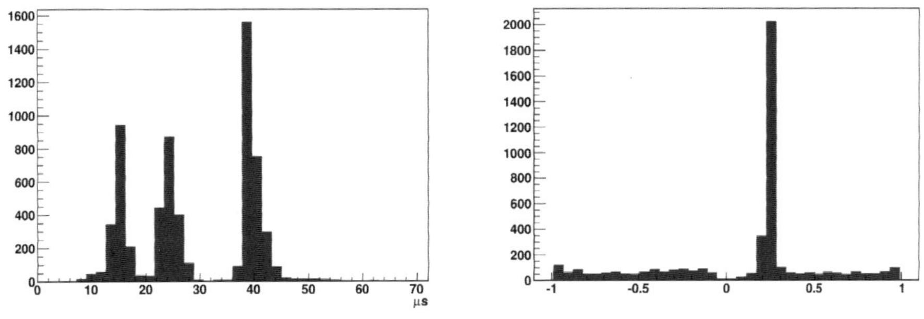

Figure 3. (L) Plasma propagation time for a cell with dust. A good cell will only have a single peak. (R) Position distribution (where 0 is at the centre of the cell, and ±1 are at the end caps) for a cell with a self-triggering point. A good cell will have a uniform distribution.

References

[1] R. Arnold *et al.*, *Nucl. Instr. Meth. A* **536** 79 (2005)

Topical Research Meeting: Prospects in Neutrino Physics (NuPhys2013) IOP Publishing
Journal of Physics: Conference Series **598** (2015) 012021 doi:10.1088/1742-6596/598/1/012021

Scattering length monitoring with the Scattering Module at the SNO+ detector

S. Langrock[1], K. Majumdar[2] for the SNO+ Collaboration

[1]School of Physics and Astronomy, Queen Mary University of London, 327 Mile End Road, London, E1 4NS, United Kingdom

[2]Department of Physics, University of Oxford, Denys Wilkinson Building, Keble Road, Oxford, OX1 3RH, United Kingdom

E-mail: s.langrock@qmul.ac.uk

Abstract. SNO+ is a multipurpose detector built to investigate neutrinoless double-beta decay and neutrino oscillations. The Scattering Module is part of the SNO+ calibration program, and is designed to measure its scattering properties using laser pulses sent into the detector at different wavelengths. We describe the modules hardware and outline the general strategy for measuring the scattering length using a cut based analysis selecting direct in-beam photons and Rayleigh scattered photons. The cut verification process for a water-filled detector is presented. The efficiency of selecting scattered photons is found to be $> 60\,\%$ with a purity in excess of $95\,\%$.

1. Introduction

The SNO+ detector is located about 2 km underground in the Creighton Mine in Sudbury, ON, Canada. Due to its location, the experiment is overburdened by a 6000 m.w.e. layer of rock [1]. Thus, background contributions from the cosmic muon flux are significantly reduced to an average of 3 muons per hour. Being the successor experiment of SNO (Sudbury Neutrino Observatory), the main structures of the detector, which are described in detail in [2], are being reused. The Acrylic Vessel (AV) with a diameter of 12 m will be filled with 780 tonnes of scintillator (Linear Alkyl Benzene or LAB). To reduce the effect of background radiation emitted from the neighbouring rock, the AV will be surrounded by 7000 tonnes of ultra pure water. Scintillation events will be registered by more than 9000 Photomultiplier Tubes (PMT). For the neutrinoless double-beta decay program, the scintillator material will be loaded with ^{130}Te.

To achieve a detailed detector profile, a sophisticated calibration program is being developed. One of its main components is the Embedded LED Light Injection Entity system, which was designed for frequent optical calibration [3]. Its optical fibres are mounted on the PMT Support structure (PSUP), enabling the measurement of different optical properties by injecting light pulses to the inner detector region without contaminating the LAB inside the AV. The Entities main calibration tasks are performed by three subsystems and consist of: timing calibration of the PMTs (Timing Module), the measurement of the Rayleigh scattering lengths in the detector (Scattering Module) and the measurement of the attenuation lengths in the detector (Attenuation Module). This paper focuses on the description of the hardware of the Scattering Module and describes the analysis strategy that will be used to measure the scattering length.

2. The Hardware of the Scattering Module

The Scattering Module is centered around four milliwatt lasers. These lasers, operating at a peak wavelength of 375, 407, 446 and 495 nm with very narrow wavelength profiles, allow the investigation of the λ^{-4} dependence of Rayleigh scattering within the detector.

Controlled via the SEPIA II Laser Driver, the lasers are connected using a bespoke Laser Switching Unit which allows fast (< 30 s) and automated changing between lasers via computer control. Slight changes in the laser output between pulses can result in a large change in the light yield and timing of the scintillator. Thus, the Module is equipped with an internal monitoring system using a high-efficiency, high-speed PMT, which records on a pulse by pulse basis. A fraction of the laser light is fed into the PMT, the output of which gives a measure of the per-pulse power and stability of the laser. Measuring the laser power will also help determining the number of photons entering the detector, which is vital for calibrating the Module internally.

Using a 5x14 mechanical-relay optical Fibre Switch, the other fraction of the laser light can be directed from any one of the four lasers to enter the detector through any one of the twelve optical fibres, leaving 2 spare fibre ports on the switch. The twelve fibres of the Scattering Module are installed in four different mounting points in the PSUP. Each of the four points houses three fibres pointing in a different direction relative to the centre of the detector: 0°, 10° and 20° (0° being directly through the centre).

3. Analysis Strategy

Only a few PMTs will be illuminated directly by the beams from the Scattering Module. Hits in the remaining PMTs will originate from scattered, absorbed and re-emitted, and reflected photons. Using timing and spatial constraints, a cut selection is applied to all photons in the Module beams in order to classify different analysis regions according to optical deflections. Figure 1 shows the time residual of all selected analysis regions. The time residual t_{res} is the time at which the PMTs have been hit, corrected by the time of flight of the photons from the fibre to the hit PMT. Hence, the earliest registered hits are direct in-beam photons which have not been obstructed in the detector, resulting in a sharp peak around $t_{\mathrm{res}} = 0$ ns. Very short time differences are observed between scattered and AV reflected light, resulting in close distributions. The observed peaks in the PSUP reflections are due to distinct PMT reflections at different angles [4].

For the measurement of the scattering properties in the detector, only the in-beam and scattered events are relevant. To determine the scattering length, the ratios of the events in these analysis regions with respect to the total number of events are calculated. These ratios will be compared between data samples taken with the Scattering Module and Monte Carlo samples simulated with different scattering lengths. Thus, the purity and efficiency of the cut selections

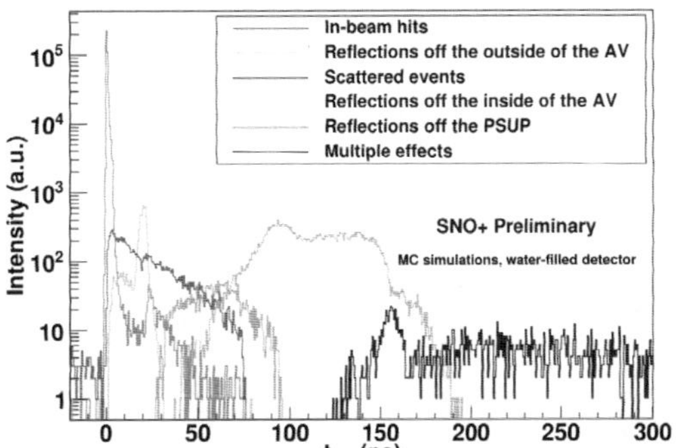

Figure 1. Time residual of events from the Scattering Module simulated with a 0° fibre and 495 nm in a water-filled detector. The time residual has been split into different analysis regions.

Table 1. Scattering cut purity and efficiency, and in-beam cut contamination with their statistic uncertainties for all wavelengths for a 0° and a 20° fibre.

	Scattering Cut		In-beam Cut
	Purity [%]	Efficiency [%]	Contamination with scattered events [%]
0° - 495 nm	96.8 ± 1.8	65.4 ± 0.7	0.6 ± 0.1
0° - 446 nm	98.2 ± 1.2	64.2 ± 0.5	0.9 ± 0.1
0° - 407 nm	98.9 ± 1.0	63.9 ± 0.4	1.3 ± 0.1
0° - 375 nm	99.2 ± 0.9	65.4 ± 0.7	1.8 ± 0.1
20° - 495 nm	96.4 ± 1.9	63.5 ± 0.7	0.5 ± 0.1
20° - 446 nm	97.8 ± 1.3	61.7 ± 0.5	0.7 ± 0.1
20° - 407 nm	98.6 ± 1.0	61.3 ± 0.4	1.0 ± 0.1
20° - 375 nm	99.0 ± 0.9	62.2 ± 0.3	1.4 ± 0.1

need to be ensured.

The tracking information for each photon from the Scattering Module was extracted from Monte Carlo samples simulating the SNO+ detector filled with H_2O. These tracks contain a history of the photons path in the detector, enabling us to know if the photon was Rayleigh scattered. The number of scattered photons in the scattering and in-beam cut was determined. To achieve the purity of the scattering cut and the contamination of the in-beam cut, these photons are compared to the total number of events selected by each cut. The efficiency of the scattering cut was calculated by comparing all scattered events to the number which has been selected by the scattering cut. Table 1 shows these values for both cuts for a selection of fibres and wavelengths.

A wavelength dependence of the Rayleigh scattering in the purity and contamination levels can be observed, resulting in higher amounts of scattering for smaller wavelengths. Over 95 % of all events selected by the scattering cut are Rayleigh scattered photons, whilst the contamination of in-beam light does not exceed 2 %. An efficiency of $(60 - 65)$ % can be achieved for the scattering cut. The efficiency values for the 20° fibre are marginally smaller due to its angled position in the detector making the applied spatial constraints less efficient.

4. Conclusion

The cut selection performs to a very high standard with very good efficiency and purity. Thus, the scattering length in the simulation can be tuned using these in-beam and scattered analysis regions to compare data events to Monte Carlo simulations.

Acknowledgements

This research has received funding from the STFC and the European Research Council under the European Union's Seventh Framework Programme (FP/2007-2013)/ERC Grant Agreement no. 278310.

References

[1] Lozza V 2012 *J. Phys.: Conf. Series* **375** 042050
[2] SNO Collaboration 2000 *Nucl. Instrum. Methods* A **449** 172–207
[3] Lefeuvre G 2013 *Nucl. Instrum. Methods* A **718** 506–8
[4] Cameron J 2001 Ph.D. thesis Oxford University

Topical Research Meeting: Prospects in Neutrino Physics (NuPhys2013) IOP Publishing
Journal of Physics: Conference Series **598** (2015) 012022 doi:10.1088/1742-6596/598/1/012022

Low background techniques for SuperNEMO

Xin Ran Liu and James Mott (for the SuperNEMO Collaboration)
Department of Physics and Astronomy, University College London, Gower Street, London WC1E 6BT, UK
E-mail: x.r.liu@ucl.ac.uk

Abstract. The UK contribution to achieving the ultra-low background conditions required inside the detectors of the SuperNEMO experiment are described. A dedicated facility has been established for the screening and selection of materials through gamma ray spectroscopy using germanium detectors. Initial results from two detectors are shown. The radon level inside the SuperNEMO detector must be less than 150 μBq/m^3 in order to achieve the target sensitivity. A Radon Concentration Line (RnCL) has been developed capable of measuring radon levels in large gas volumes down to 5 μBq/m^3, improving on standard state-of-the-art radon detectors by 3 orders of magnitude. The development, commissioning and first measurements of radon content using the RnCL are also presented.

1. Introduction

SuperNEMO [1] is a next generation neutrinoless double beta decay experiment based on the successful NEMO-3 tracker-calorimeter design [2]. The main experimental objective is to achieve a half-life sensitivity of 10^{26} years corresponding to an effective Majorana neutrino mass of $\langle m_{\beta\beta}\rangle$ < 50 - 100 meV.

One of the main challenges in achieving the desired sensitivity is background from natural radioactivity most importantly coming from ^{238}U and ^{232}Th decay chains. The most critical of these backgrounds comes from the decay daughters of radon, ^{222}Rn and ^{220}Rn, ^{214}Bi (Q_β= 3.27 MeV) and ^{208}Tl (Q_β= 4.99 MeV) respectively, which have enough energy to mimic a double-β event.

An additional problem comes from the fact that radon can enter the detector either through diffusion, contamination during detector construction or emanation from the detector materials themselves. The main problem comes from the isotope of ^{222}Rn due to its relatively long half-life of 3.8 days. As a result, all construction materials are screened to ensure radiopurity, and materials in direct contact with the tracker gas are measured for ^{222}Rn emanation.

The first SuperNEMO detector module, the Demonstrator, is under construction. The most stringent radiopurity requirements, A(^{214}Bi) < 10 μBq/kg, A(^{208}Tl) < 2 μBq/kg, apply for the source foil of the $\beta\beta$ isotope of choice, ^{82}Se. This sensitivity is reached with a dedicated detector described in [3]. For the other detector materials the radiopurity requirements are at a level of 0.1 - 10 mBq/kg depending on its position inside the detector. These materials are selected using low background Germanium detector gamma-spectroscopy. In addition, radon emanation from detector sub-modules and ^{222}Rn concentration in the tracking detector gas are measured at a level of < 0.15 mBq/m^3, the level required to achieve the SuperNEMO design sensitivity.

Here we report on the last two types of radiopurity measurements (Germanium spectroscopy and radon emanation).

2. Boulby Underground laboratory

Boulby is a working potash and rock salt mine, located in the North East of England, the deepest in the UK, at 1070 metres (2805 m w.e.) [4]. The lab is built inside the mined tunnel surrounded by salt which is very low in natural radioactive backgrounds.

2.1. Germanium detectors

So far two detectors have been commissioned for material screening at Boulby. One is an Ortec coaxial low-background high purity germanium (HPGe) detector and the other is a Canberra broad energy germanium (BEGe) detector with sensitive masses of 1.9 kg and 0.5 kg respectively. The detectors are surrounded by passive shielding made of lead and OFHC copper. An anti-radon purging system was constructed for the Ortec HPGe detector which allowed a factor of 4 reduction from radon related backgrounds. The background spectra of the detectors are shown in Fig. 1.

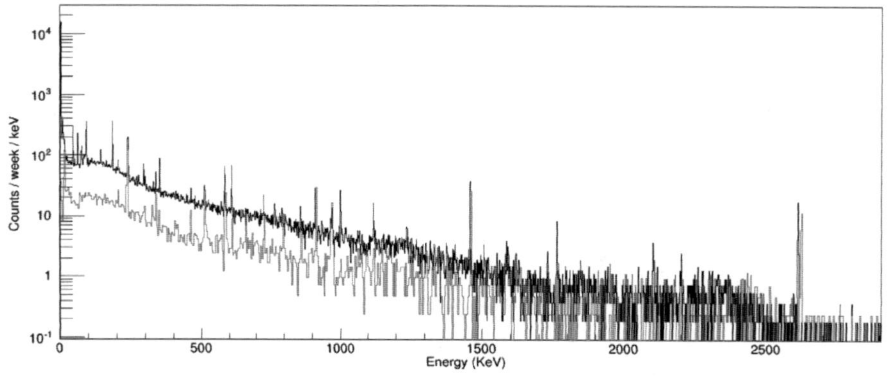

Figure 1. Background spectra of Ortec HPGe (red) and Canberra BEGe (black).

The integral counting rate of the HPGe and BEGe detectors in the 100 - 2700 keV energy region are 0.768 ± 0.001 and 2.381 ± 0.001 respectively. In the case of the BEGe detector the shielding was non-optimised and is currently undergoing a major refurbishment. Nonetheless, both detectors were able to achieve a sensitivity to ^{238}U and ^{232}Th progenies at a level of 1-3 mBq/kg after 3 weeks of measurement, depending on the geometry and composition of the measured sample.

2.2. Detector upgrades

Both the coaxial and the BEGe detector will undergo upgrades to further improve performance. These include conversion of the Ge-crystal cooling system to a J-type configuration to minimise a direct line of sight gamma-ray background and having customised shielding tailored to further suppress backgrounds. In addition, the coaxial detector will have a larger LN_2 dewar and the magnesium endcap will be replaced by one of carbon fibre to improve radiopurity and increase detection efficiency at lower energies. Both detectors will have an upgraded anti-Rn purging system.

Journal of Physics: Conference Series **598** (2015) 012022 doi:10.1088/1742-6596/598/1/012022

3. Low-Level Radon Detector

An electrostatic detector similar to the one described in [5] was used for radon detection. It collects radon progenies on a silicon PIN diode using an electric field. These are then identified by the energy they deposit when undergoing an α-decay. Isotopes of ^{218}Po and ^{214}Po can clearly be seen in an energy spectrum from the detector as shown in Fig. 2.

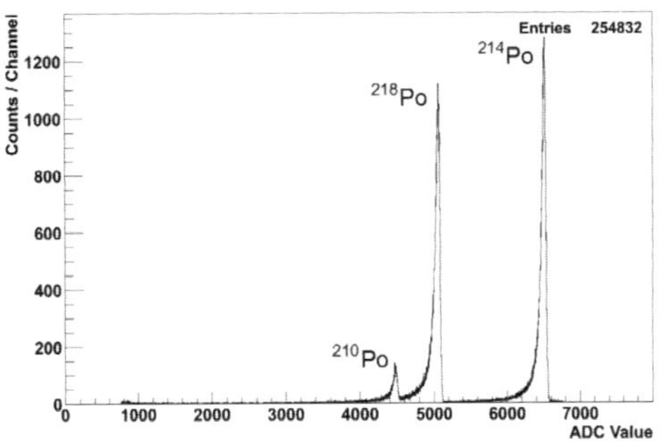

Figure 2. Energy spectrum from a radon source.

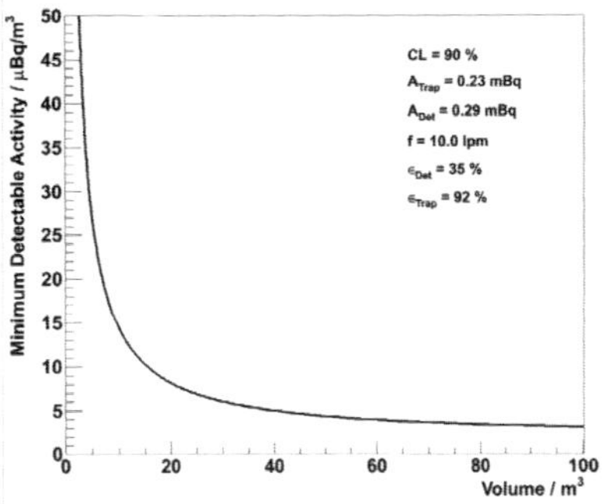

Figure 3. Sensitivity estimates of the RnCL with helium.

The detection efficiency depends on the isotope and the carrier gas. The highest efficiency of 35% was achieved with ^{214}Po in helium. With a background of 7$\pm$1 counts per day the detector's sensitivity to radon levels is 1.7 mBq/m^3 (90% CL).

3.1. Radon Concentration Line

Electrostatic detectors are typically sensitive to 1 mBq/m^3 which is higher than the SuperNEMO target radon concentration of 0.15 mBq/m^3. In order to measure detector sub-modules to this level during construction a Radon Concentration Line (RnCL) has been developed.

The concept behind the RnCL is to "concentrate" Rn by pumping a large volume of gas through a cold ultra-pure activated-carbon trap where Rn is adsorbed. The trap is then heated and this concentrated radon is transferred into an electrostatic detector via helium purge.

3.2. Sensitivity and First Measurements

The trapping and transfer efficiency of the radon concentration line has been calibrated using a dedicated ^{222}Rn source and found to be as high as 92% with helium as a carrier gas. This translates into sensitivity levels for radon concentration down to O(10) μBq/m^3 for large gas volumes as shown in Fig 3. The first measurements of commercially available helium and nitrogen gases have now been made and are shown in table 1.

Table 1. Result from measurement of radon present in helium and nitrogen gas.

Gas	Source	Radon Level (μBq/m^3)
He	Cylinder	70-100
N_2	Cylinder	400-1000
N_2	Boil-off	90-140

One interesting observation was that the nitrogen gas in cylinders seems to have a systematically higher content of ^{222}Rn compared to helium although improvement is seen with gas from boil-off from liquid N_2. Based on these results a dedicated facility has been built to clean the SuperNEMO working gas before it enters the tracking detector.

4. Conclusions

A low background screening facility has been established at the Boulby Underground Laboratory. Two Ge-detectors have been commissioned to carry out material screening campaigns for SuperNEMO and other low background experiments with a sensitivity of 1 mBq/kg and better. In addition, an ultra-sensitive ^{222}Rn detection facility has been commissioned with a sensitivity down to 5 μBq/m^3 when large volumes of gas are sampled. The facility will be also used to measure ^{222}Rn emanation from SuperNEMO detector sub-modules during construction with a sensitivity below 0.15 mBq/m^3.

5. Acknowledgments

The authors would like to thank Emma Meehan and Sean Paling of Boulby Underground Laboratory for their support and assistance.

References

[1] Arnold R *et al* 2010 *Eur. Phys. J.* C **70** 927-43
[2] Arnold R *et al* 2005 *Nucl. Instrum. Meth.* A **536** 79-122
[3] Argyriades J *et al* 2010 *Nucl. Instrum. Meth.* A **622** 120-28
[4] Robinson M *et al* 2003 *Nucl. Instrum. Meth.* A **511** 347-53
[5] Choi E *et al* 2001 *Nucl. Instrum. Meth.* A **459** 177-81

Rejection of backgrounds from pileup alpha coincidences in the SNO+ detector

K Majumdar for the SNO+ Collaboration

Department of Physics, University of Oxford, Denys Wilkinson Building, Keble Road, Oxford, OX1 3RH, United Kingdom

E-mail: Krishanu.Majumdar@physics.ox.ac.uk

Abstract. SNO+ is a multi-purpose neutrino detector with a number of physics goals, one of which is to investigate the existence of Neutrinoless Double Beta Decay. The potential rarity of this process means that other radioactive decays occurring in and around the detector can become major background sources. Two such backgrounds are the β decays of ^{214}Bi and ^{212}Bi, which are followed by the α decays of their respective daughters: ^{214}Po and ^{212}Po. These daughters have relatively short half-lives, and so these β-α coincidences can form pileup events in the detector. Two methods are presented here for rejecting such pileup coincidence events.

1. Introduction

SNO+ is the successor to the Sudbury Neutrino Observatory (SNO), located 2km underground in Ontario, Canada. It consists of a 6-metre radius spherical volume of liquid scintillator surrounded by 7000 tonnes of ultra-pure water and observed by over 9000 photomultiplier tubes (PMTs) that detect the light from scintillation events (a more detailed description of the experiment is given in [1]). The detector's location, with a 6000 m.w.e. rock overburden [2], provides shielding from cosmic particle flux, and the ultra-pure water reduces the number of backgrounds entering from the surrounding rock. However, there are still backgrounds present in the scintillator itself, including the β decays of ^{214}Bi and ^{212}Bi, which are followed by the α decays of their respective daughters ^{214}Po and ^{212}Po. This β-α coincidence can occur completely within a single 400ns-wide trigger window due to the short half-lives of the Po daughters, making the coincidence a pileup event that appears very similar to a potential ^{130}Te neutrinoless double beta decay (ββ0ν) event.

The time residual, t^i_{res} of a single triggered PMT in any given event is calculated by:

$$t^i_{\text{res}} = t^i_{\text{pmt}} - t^i_{\text{flight}} - t_{\text{ev}}$$

where t^i_{pmt} is the ith PMT's trigger time, t_{ev} is the reconstructed event time, and t^i_{flight} is the travel time through the varying materials between the event and PMT. Figure 1 shows the normalised t_{res} distributions for ^{130}Te ββ0ν, ^{214}Bi β and ^{214}Po α events (the distributions for ^{212}Bi and ^{212}Po look identical to their 214- counterparts).

2. Cumulative Time Residuals Method

The cumulative number of triggered PMTs up to a given value of t_{res} is a reliable way of viewing the differences between ^{130}Te ββ0ν and BiPo events. Figure 2 shows the cumulative fraction of triggered

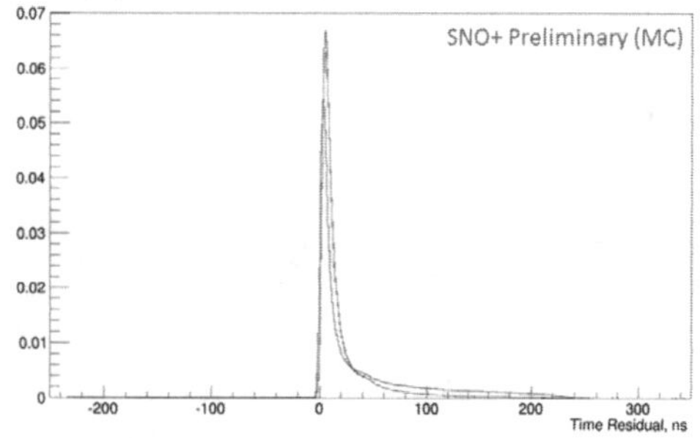

Figure 1. The normalised t_{res} distributions for ^{130}Te ββ0ν, ^{214}Bi β, and ^{214}Po α events. The Te and Bi distributions are almost identical, but the Po distribution has a lower peak and longer tail, both resulting from the larger late-light component associated with scintillating α's.

PMTs as a function of t_{res} for a typical BiPo pileup event (the fraction is used rather than the raw number of PMTs in order to correct for the slight variation in the raw number of PMTs per event).

A Kolmogorov-Smirnov test between the BiPo and CDF curves quantifies the difference, using:

$$\Gamma = \frac{1}{Nhits} \times \sum_b \left(x_b^{\text{ev}} - x_b^{\text{cdf}}\right)^2$$

where there are *b* bins in each distribution, and x_b is the normalized number of events in bin *b*. Figure 3 shows the distribution of Γ for ^{130}Te ββ0ν and ^{214}BiPo events.

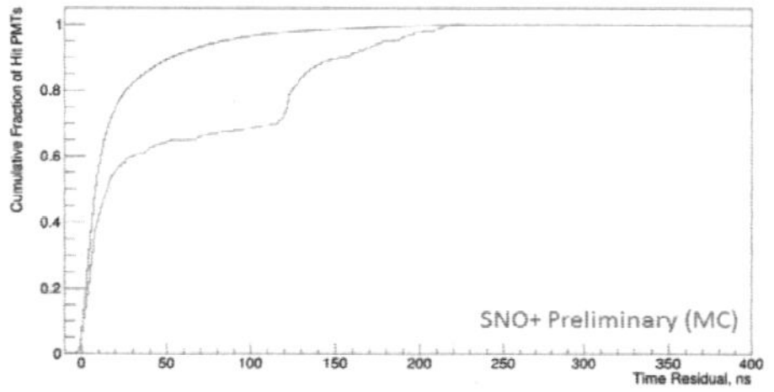

Figure 2. The normalised cumulative fraction of triggered PMTs as a function of t_{res} for ^{130}Te ββ0ν events (smooth curve, top) and a BiPo pileup event (double-shouldered curve, bottom).

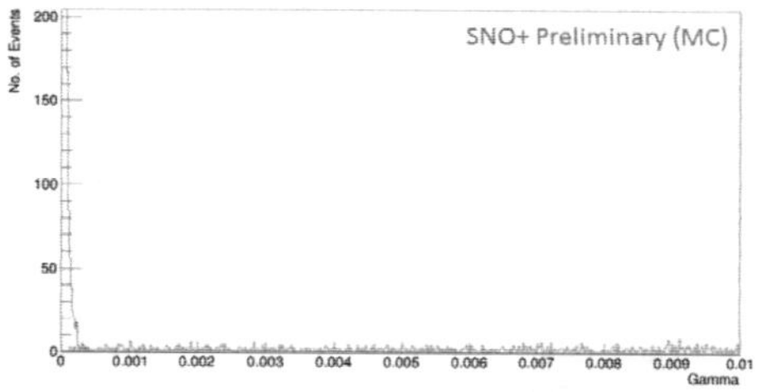

Figure 3. The values of Γ for ^{130}Te ββ0ν (left-side peak) and ^{214}BiPo (low-level spread) events. ^{212}BiPo events have an almost identical distribution to the latter. There is a very clear separation between the Te and BiPo events.

3. Log-Likelihood Difference Method

The log-likelihood of an event being ^{130}Te ββ0ν based on its t_{res} distribution is given by:

$$\mathcal{L}_{\text{Te}} = \sum_{i=1}^{N} \ln\left(N \text{P}_{\text{Te}}\left(t_{\text{res}}^{i}\right)\right)$$

where *N* is the number of triggered PMTs and $\text{P}_{\text{Te}}\left(t_{\text{res}}^{i}\right)$ is the probability of the ith PMT having the time residual t_{res}^{i} based on the ^{130}Te t_{res} distribution in Figure 1. Similarly, using the Bi and Po t_{res} distributions, the log-likelihood of the same event being a BiPo pileup is:

$$\mathcal{L}_{\mathrm{BiPo}} = \sum_{\mathrm{i=1}}^{\mathrm{N}} \ln\left([N - A]\mathrm{P}_{\mathrm{Bi}}(t^{i}_{\mathrm{res}}) + A\mathrm{P}_{\mathrm{Po}}(t^{i}_{\mathrm{res}} - \Delta t)\right)$$

where A is the mean number of triggered PMTs in a Po α event (different for ^{214}Po and ^{212}Po). The quantity Δt represents the delay before the α emission, and is varied to maximize $\mathcal{L}_{BiPo}$. This leads to a minimised difference:

$$\Delta\mathcal{L} = \mathcal{L}_{\mathrm{Te}} - \mathcal{L}_{\mathrm{BiPo}}$$

Figure 4 shows the distribution of $\Delta\mathcal{L}$ for ^{130}Te ββ0ν and ^{214}BiPo pileup events.

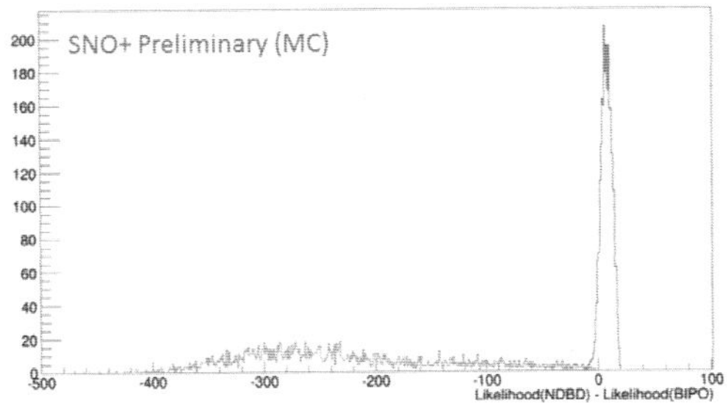

Figure 4. The values of $\Delta\mathcal{L}$ for ^{130}Te ββ0ν (right-side peak) and ^{214}BiPo (low-level spread) events. ^{212}BiPo events have an almost identical distribution to the latter. There is a very clear separation between the Te and BiPo events.

4. Overall Bi-Po Rejection

Figure 5 shows the remaining BiPo pileup percentage as a function of the remaining ^{130}Te ββ0ν percentage, where each point corresponds to a cut-value on either Γ or $\Delta\mathcal{L}$. (The ^{212}BiPo curves are limited by statistics.)

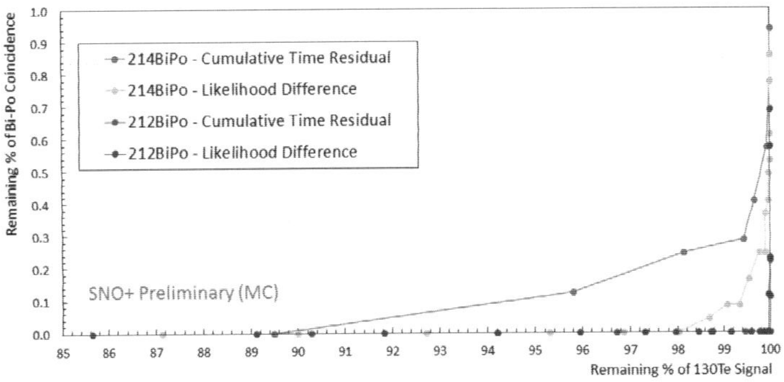

Figure 5. The remaining percentage of BiPo pileup events as a function of the remaining percentage of 130Te ββ0ν events, for the four method/isotope combinations that have been considered.

In both backgrounds, the Log-Likelihood Difference method performs better than the Cumulative t_{res} method. However, there are still a small number of BiPo events remaining. These events all have a very fast α, i.e. one that has been emitted <15ns after the Bi β. In this situation, the BiPo t_{res} distribution looks identical to that of ^{130}Te ββ0ν, making discrimination impossible.

References

[1] SNO Collaboration 2000, *Nucl. Instrum. Methods* A **449** 172-207

[2] Lozza V 2012 *J. Phys.: Conf. Series* **375** 042050

Monitoring nuclear reactors for safeguards purposes using anti-neutrinos

J Carroll, J Coleman, M Lockwood, C Metelko, M Murdoch*, C Touramanis
Oliver Lodge Laboratory, Oxford Street, University of Liverpool, Liverpool, L69 7ZE

G Davies, A Roberts
Wylfa Power Station, MagnoxSites, Angelsey, UK

Abstract. Preventing nuclear proliferation is a high priority for the international community. Monitoring of nuclear facilities to detect unauthorised removal of fissile materials from operational cores is central to this. Neutrino detection devices can be used to remotely monitor the core of operating reactors in a safe, reliable manner. Technology developed for the T2K experiment can be adapted to make a small footprint, reliable, anti-neutrino detector. Through, characterisation of the anti-neutrino spectrum there is a possibility to provide core material accountancy.
A prototype of such a device has been developed and demonstrated at the University of Liverpool. Based on the design of the T2K Near Detector Calorimeter, the device will detect anti-neutrinos through the distinctive delayed coincidence signal of inverse beta decay interactions. This poster presented data from detector commissioning. The detector is currently deployed at Wylfa power station, UK for field testing.

1. Anti-neutrino Reactor Monitoring for Safeguards

Safeguarding of nuclear materials is a concern for the global community. The detection of unauthorized production and diversion of weapon's grade materials such as Plutonium-239 is particularly important to this end. Plutonium accounting is difficult for several 4th generation reactor designs such as pebble bed and liquid core reactors that do not have itemised fuel content.

Reactor anti-neutrino detection has been used in several neutrino oscillation experiments providing measurements of the mixing angles θ_{12}[1] and θ_{13}[2]. At reactor neutrino energies cross-section of inverse beta decay interaction, $\bar{\nu}_e + p \rightarrow e^+ + n$, is dominant. The detection of prompt energy deposit from the positron and the delayed energy despoit from the neutron capture gives a delayed conicidence signal for enhanced background rejection. It is theoretically possible to unfold a reactor's anti-neutrino flux and spectra to measure the plutonium content of the core in near to real time. As such, anti-neutrino detectors could provide material accountance in a manner that is non-intrusive to reactor systems. The anti-neutrino flux cannot be effectively shielded to obscure covert operations or material diversion. These qualities can potentially make anti-neutrino based safeguards an attractive prospect with particular advantages for use with non-itemised reactor designs.

The detectors used in oscillation experiments are large (~1 kiloton) and expensive making them unsuitable for safeuards usage. Safeguards detectors must be of order 1 ton in order to be placed in close proximity to a reactor whilst being cheap enough for wide scale deployment. Lawrence Livermore National Laboratory developed a prototype, liquid scintillator based, anti-neutrino detector of order 1 ton in size. The detector was deployed at the San Onofre Generating Station where reactor anti-neutrinos were used to observe both the reactor's operational state (on/off) and the fuel burn up during operations [3] as shown in Figure 1.

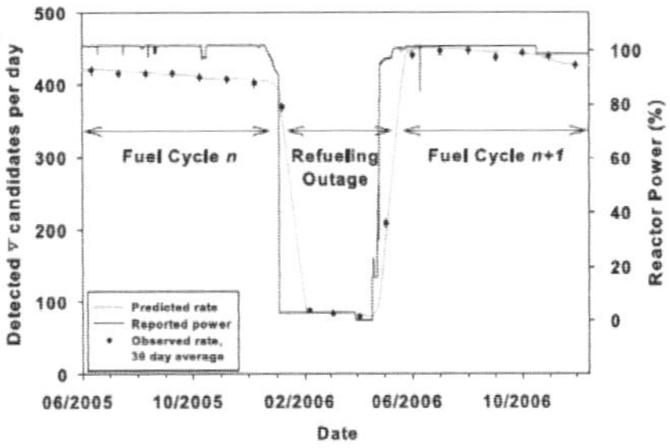

Figure 1 Anti-neutrino event rate in the SONGS1 detector showing fuel burn-up and refuelling period [3]

The observations made by the SONGS1 detector successfully showed the concept of a small scale reactor anti-neutrino monitoring device to be sound. However, the use of flammable liquid scintillator is not ideal for safeguards usage where the detector must be located close to the reactor core. Any ideal safeguards detector should also be robust, transportable and easy to operate. Failing to adhere to these ideal criteria creates an obstacle to use and wide scale deployment.

2. A Reactor Monitoring Detector based on the T2K ECal

The electromagnetic calorimeter (ECal) [4] for the T2K [5] near detector is a sampling calorimeter using plastic scintillator active layers with lead conversion layers. This technology suits many of the criteria for an ideal safeguards detector. The ECal modules were constructed at Liverpool University and Daresbury laboratory before being shipped to the J-PARC facility in Japan. The plastic scintillator poses little fire or explosion risk, is extremely robust and cheap. Robust and cheap Hammamatsu Multi-Pixel Photon Counters (MPPCs) coupled to wavelength shifting fibers are used for optical readout of the scintillator.

The calorimeter design has been modified using a gadolinium compound to act as a neutron capture agent, converting it into an anti-neutrino detector design. Gadolinium has a large thermal neutron capture cross-section (250,000 barns) and emits an 8 MeV gamma ray cascade giving a distinct signal for effective background rejection.

The fully constructed anti-neutrino detector contains more than 1 ton of active plastic scintillator. A picture of the full detector is shown in Figure 2. An internal veto system has been designed for rejection of cosmic ray events and to allow effective operation with no overburden, i.e. above ground. T2K Readout Merger Module (RMM) [5] boards have been adapted to accept inverse beta decay triggers as well as the cosmic ray veto signal.

Figure 2 A picture of the completed anti-neutrino detector

3. Detector Commissioning and Testing

A small scale prototype was built with removable gadolinium dopant to observe the difference between gadolinium and hydrogen neutron captures (hydrogen has a smaller capture cross section and gives a single 2.2 MeV gamma ray on neutron capture). Californium-252 was used as a neutron source. The source was enclosed in lead shielding to attenuate out its gamma ray emissions. Figure 3 shows the clear difference in neutron capture rates with and without gadolinium.

A full scale detector was subsequently constructed and tested using the same californium source with shielding. The full scale detector was also exposed to a colbalt-60 gamma ray source. Figure 4 shows the clear neutron-gamma discrimination of the detector.

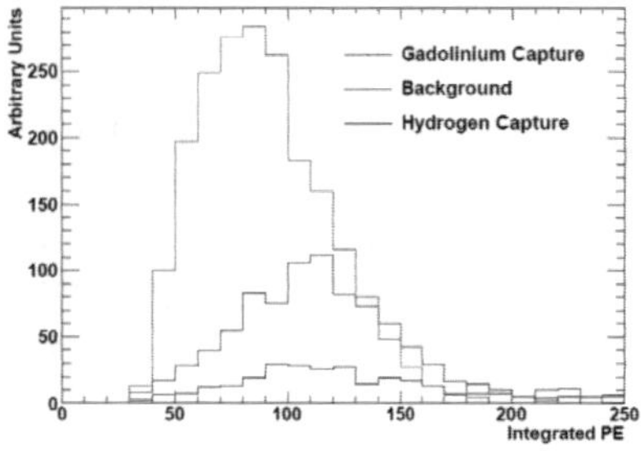

Figure 3 A histogram of integrated charge deposit for neutron capture events in a small prototype detector. The integrated area of the curves is proportional to the neutron capture rate

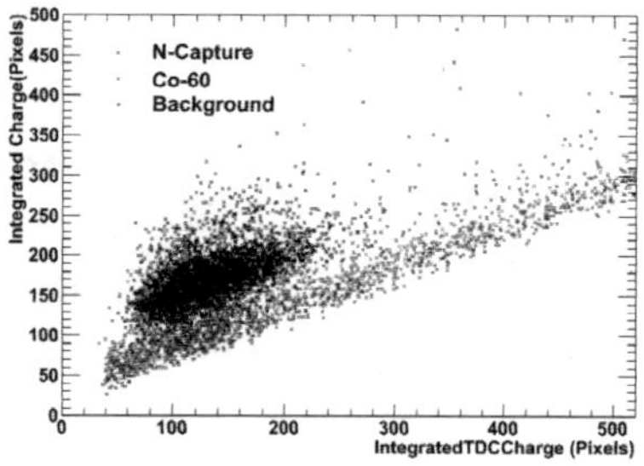

Figure 4 The neutron-gamma discrimination abilities for the full scale detector using Cf-252 and Co-60 sources. The x axis shows integrated charge for individual MPPC hits. The y axis shows total integrated charge.

4. Summary

An anti-neutrino detector for safeguards purposes has been constructed and commissioned at Liverpool based on the technology used for the T2K near detector ECal. Commissioning data shows the detector to be highly effective at neutron-gamma ray discrimination due to the use of a gadolinium neutron absorber. Combined with the ECal's ability to detect positrons, the anti-neutrino detector is well suited to the observation of inverse beta decay interactions. The detector has been deployed at the Wylfa reactor site, Anglesey, UK for detector demonstration in a safeguards environment and is currently awaiting reactor start-up.

Acknowledgements

The authors acknowledge the support of the STFC and the Royal Society. The authors are grateful for the support and contributions of the Particle Physics group and the mechanical workshop from the Physics Department at the University of Liverpool, and Support of the Site and Staff of Wylfa Power Station.

References

[1] Eguchi, K *et al* 2003 First results from KamLAND: Evidence for reactor anti-neutrino disappearance *Phys.Rev.Lett* **90** 021802

[2] An, F.P *et al* 2012 Observation of electron-antineutrino disappearance at Daya Bay *Phys.Rev.Lett* **108** 171803

[3] Bowden, N S 2008 Reactor Monitoring and safeguards using antineutrino detectors *J.Phys.Conf.Ser.* **136** 022008

[4] Allan D *et al* 2013 The electromagnetic calorimeter for the T2K near detector ND280 *JInst* **8** 10019

[5] Abe K *et al* The T2K Experiment *Nuclear Instruments and Methods A* **659** 106-135

Neutrino Mixing from $\Delta(6n^2)$ Groups

Thomas Neder [1]*

[1] *School of Physics and Astronomy, University of Southampton, Southampton, SO17 1BJ, U.K.*

*E-mail: T.Neder@soton.ac.uk

Abstract

Experimentally viable lepton mixing parameters can be predicted in so-called direct flavour models with Majorana neutrinos using $\Delta(6n^2)$ groups as a flavour group. In direct models, in which the flavour group is broken to a $Z_2 \times Z_2$ subgroup in the neutrino sector, mixing angles and Dirac CP phase are purely predicted from symmetry. General predictions of direct models with $\Delta(6n^2)$ flavour groups are that all mixing angles are fixed up to a discrete choice and that the Dirac CP phase is 0 or π; Furthermore, the middle column of the mixing matrix is trimaximal which yields the sum rule $\theta_{23} = 45^\circ \mp \theta_{13}/\sqrt{2}$ depending on the Dirac phase. These predictions of lepton mixing parameters are compatible with recent global fit results or will be tested experimentally in the near future. It is the first time that such predictions have been obtained model-independently for an infinite series of groups.

Introduction

The problem of the origin of neutrino masses and mixing is of fundamental importance and among the more specific questions that models of neutrinos have to attempt to answer are:

- Why do particles mix as they do?
- Why are there 3 families of particles?
- Why are the masses of the particles what they are?
- Is there CP violation in the lepton sector?

A succesful and numerous class of models of neutrino masses and mixings are models with flavour symmetries: Generations of fermions are assigned to representations of an additional symmetry group called the flavour symmetry; for a review see e.g. [1]. With this additional symmetry, the full symmetry of the model becomes $G_{\text{SM}} \times G_{\text{Flavour}}$. The additional invariance of the Lagrangian under G_{Flavour} restricts the allowed couplings in the Yukawa sector and eventually the allowed mass and mixing matrices. Often, the flavour symmetry needs to be broken, because the maximal flavour symmetry under which a Majorana mass term for three neutrinos with non-degenerate masses can be invariant is $Z_2 \times Z_2$. This breaking can e.g. be achieved using scalar fields.

If the the flavour group is both broken to a $Z_2 \times Z_2$ subgroup in the neutrino sector of the theory and to a discrete subgroup, often a cyclic group, in the sector of charged leptons, a model with a flavour symmetry is called *direct*. In a direct model, the mixing angles and the Dirac CP phase are purely predicted from the choice of flavour group and its unbroken subgroups.

Many promising flavour symmetry candidates are $\Delta(6n^2)$ groups [2, 3, 4, 5, 6, 7, 8, 9, 10]. In particular, after the measurement of θ_{13} [11, 12, 13], the only flavour symmetries in direct models that remain viable are $\Delta(6n^2)$ groups or subgroups thereof [14, 15]. Furthermore, in direct flavour models, one can analyse $\Delta(6n^2)$ groups for all even n simultaneously [16].

The groups $\Delta(6n^2)$ are non-abelian groups of order $6n^2$ and are isomorphic to a semidirect product: $\Delta(6n^2) \cong (Z_n \times Z_n) \rtimes S_3$ [17]. The maybe best-known members of the series are $S_3 = \Delta(6)$ and $S_4 = \Delta(24)$.

Any $Z_2 \times Z_2$ subgroup of every $\Delta(6n^2)$ can give rise to a different mixing matrix. Consequently, all mixing matrices that are allowed by a certain $\Delta(6n^2)$ group can be found systematically by listing all $Z_2 \times Z_2$ subgroups for each $\Delta(6n^2)$ group. In direct models with a $\Delta(6n^2)$ flavour group, only breaking $\Delta(6n^2)$ to Z_3 in the sector of charged leptons will produce experimentally viable predictions for mixing angles [16].

To produce concrete results for the mixing matrix, left-handed doublets are to transform (without loss of generality) under a 3-dimensional representation with the generators (where $\eta = e^{2\pi i/n}$):

$$a = \begin{pmatrix} 0 & 1 & 0 \\ 0 & 0 & 1 \\ 1 & 0 & 0 \end{pmatrix}, b = -\begin{pmatrix} 0 & 0 & 1 \\ 0 & 1 & 0 \\ 1 & 0 & 0 \end{pmatrix}, c = \begin{pmatrix} \eta & 0 & 0 \\ 0 & \eta^{-1} & 0 \\ 0 & 0 & 1 \end{pmatrix}, d = \begin{pmatrix} 1 & 0 & 0 \\ 0 & \eta & 0 \\ 0 & 0 & \eta^{-1} \end{pmatrix}. \tag{1}$$

All $Z_2 \times Z_2$ subgroups of a $\Delta(6n^2)$ group can be indexed by a single number which takes values $\gamma = 1, \ldots, n/2$ [16]. The columns of the mixing matrix are now given by the common eigenvectors of the elements of each $Z_2 \times Z_2$ subgroup in the basis where the mass matrix of the charged leptons is already diagonal. Note that this approach does not determine the ordering of the columns and rows of the mixing matrix. The ordering of the mixing matrix can be fixed by demanding that the smallest entry be in the top right of the mixing matrix, i.e. V_{13}. With $\vartheta = \pi\gamma/n$ every mixing matrix has up to ordering the following form:

$$V = \begin{pmatrix} \sqrt{\frac{2}{3}}\cos(\vartheta) & \frac{1}{\sqrt{3}} & \sqrt{\frac{2}{3}}\sin(\vartheta) \\ -\sqrt{\frac{2}{3}}\sin\left(\frac{\pi}{6}+\vartheta\right) & \frac{1}{\sqrt{3}} & \sqrt{\frac{2}{3}}\cos\left(\frac{\pi}{6}+\vartheta\right) \\ \sqrt{\frac{2}{3}}\sin\left(\frac{\pi}{6}-\vartheta\right) & -\frac{1}{\sqrt{3}} & \sqrt{\frac{2}{3}}\cos\left(\frac{\pi}{6}-\vartheta\right) \end{pmatrix}. \tag{2}$$

This lepton mixing matrix is trimaximal with θ_{13} fixed up to a discrete choice and CP-phase 0 or π. For any value of ϑ the Dirac CP phase can be both 0 or π which corresponds to interchanging the two rows that are not fixed because they do not contain the smallest entry. This will also result in a different value for θ_{23}. Generally, if the middle column is trimaximal, the following sum rule holds: $\theta_{23} = 45^\circ \mp \theta_{13}/\sqrt{2}$ [1].

The allowed values of $|V_{13}|$ are plotted in figure (1), where every choice of γ and n produces a dot. The number of allowed $Z_2 \times Z_2$ subgroups and hence the number of allowed mixing matrices increases with n. If n can be divided by 3, some of the subgroups produce the same mixing matrices which explains when dots seem to be missing. The horizontal lines denote the present approximate 3σ range of $|V_{13}|$ from [18].

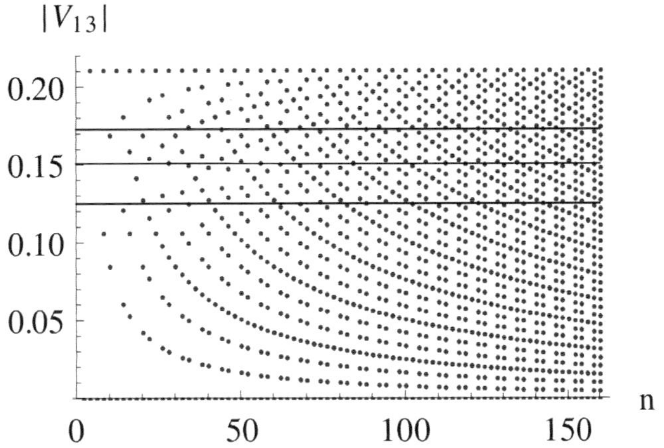

Figure 1: Possible values of $|V_{13}|$; the lines denote the present approximate 3σ range of $|V_{13}|$ from [18]. Examples include $|V_{13}| = 0.211, 0.170, 0.160, 0.154$ for $n = 4, 10, 16, 22$, respectively. Each value of V_{13} allows for two values of θ_{23} with $\delta_{CP} = 0$ and $\delta_{CP} = \pi$ given by $\theta_{23} = 45° \mp \theta_{13}/\sqrt{2}$ respectively.

Conclusions and Outlook

Direct models with a $\Delta(6n^2)$ flavour symmetry for Majorana neutrinos predict experimentally viable lepton mixing matrices. In [16] lepton mixing predictions from $\Delta(6n^2)$ have been obtained for arbitrary even values of n and all mixing matrices are shown to have the form of Eq. (2). The predictions are compatible with current results of global fits and will be further tested in the near future.

A direct model only predicts mixing angles and the Dirac CP phase from symmetry while Majorana phases remain unconstrained. Majorana phases can be predicted from symmetry in models with a generalised CP symmetry. For direct models with $\Delta(6n^2)$ flavour groups, generalised CP symmetries have been studied in [19].

Acknowledgements

TN acknowledges partial support from the European Union FP7 ITN-INVISIBLES (Marie Curie Actions, PITN- GA-2011- 289442).

Journal of Physics: Conference Series **598** (2015) 012025 doi:10.1088/1742-6596/598/1/012025

References

[1] S. F. King and C. Luhn, Rept. Prog. Phys. **76** (2013) 056201 [arXiv:1301.1340 [hep-ph]]; S. F. King, A. Merle, S. Morisi, Y. Shimizu and M. Tanimoto, arXiv:1402.4271 [hep-ph].

[2] R. d. A. Toorop, F. Feruglio and C. Hagedorn, Phys. Lett. B **703**, 447 (2011). [arXiv:1107.3486 [hep-ph]].

[3] R. de Adelhart Toorop, F. Feruglio and C. Hagedorn, Nucl. Phys. B **858**, 437 (2012). [arXiv:1112.1340 [hep-ph]].

[4] G. -J. Ding, Nucl. Phys. B **862**, 1 (2012). [arXiv:1201.3279 [hep-ph]].

[5] S. F. King, C. Luhn and A. J. Stuart, Nucl. Phys. B **867**, 203 (2013). [arXiv:1207.5741 [hep-ph]].

[6] C. S. Lam, Phys. Rev. D **87**, no. 5, 053012 (2013). [arXiv:1301.1736 [hep-ph]].

[7] S. F. King and C. Luhn, JHEP **0910**, 093 (2009). [arXiv:0908.1897 [hep-ph]].

[8] I. de Medeiros Varzielas and G. G. Ross, JHEP **1212**, 041 (2012). [arXiv:1203.6636 [hep-ph]].

[9] R. Krishnan, J. Phys. Conf. Ser. **447**, 012043 (2013). [arXiv:1211.3364 [hep-ph]].

[10] M. Holthausen and K. S. Lim, Phys. Rev. D **88**, 033018 (2013). [arXiv:1306.4356 [hep-ph]].

[11] F. P. An *et al.* [DAYA-BAY Collaboration], Phys. Rev. Lett. **108** (2012) 171803 [arXiv:1203.1669];

[12] J. K. Ahn *et al.* [RENO Collaboration], Phys. Rev. Lett. **108** (2012) 191802 [arXiv:1204.0626].

[13] Y. Abe *et al.* [Double Chooz Collaboration], arXiv:1301.2948 [hep-ex].

[14] C. S. Lam, Phys. Rev. D **87** (2013) 013001 [arXiv:1208.5527 [hep-ph]].

[15] M. Holthausen, K. S. Lim and M. Lindner, Phys. Lett. B **721** (2013) 61 [arXiv:1212.2411 [hep-ph]].

[16] S. F. King, T. Neder and A. J. Stuart, Phys. Lett. B **726**, 312 (2013). [arXiv:1305.3200 [hep-ph]].

[17] J. A. Escobar and C. Luhn, J. Math. Phys. **50** (2009) 013524 [arXiv:0809.0639 [hep-th]].

[18] F. Capozzi, G. L. Fogli, E. Lisi, A. Marrone, D. Montanino and A. Palazzo, [arXiv:1312.2878 [hep-ph]].

[19] S. F. King and T. Neder, [arXiv:1403.1758 [hep-ph]].

Journal of Physics: Conference Series **598** (2015) 012026 doi:10.1088/1742-6596/598/1/012026

Atmospheric neutrino oscillations with PINGU

Sirin Odrowski for the IceCube-PINGU collaboration[1]
Dept. of Physics, University of Alberta, Edmonton, Alberta, Canada T6G 2E1
E-mail: Sirin.Odrowski@icecube.wisc.edu

Abstract. Measurements of atmospheric, solar, reactor and accelerator neutrinos (see e.g. [1, 2, 3, 4]) have shown that neutrinos undergo flavour-changing oscillations and hence have mass. The cubic-kilometre neutrino detector IceCube [5] with its more densely instrumented inner region DeepCore [6] has studied oscillations of atmospheric neutrinos at energies above $\sim$ 10 GeV [7, 8, 9]. The Precision IceCube Next Generation Upgrade (PINGU) [10] is a proposed extension of IceCube with the goal to perform precise measurements of atmospheric neutrino oscillations down to a few GeV and to determine the neutrino mass hierarchy (NMH).

1. Introduction

Neutrino oscillations can only exist if neutrinos have mass, and the probability to change flavour depends on the squared mass difference and the mixing angles. The ordering of the neutrino masses is however not fully resolved yet as the sign of Δm^2_{atm} is still unknown. The third neutrino mass eigenstate may thus either be the heaviest (normal mass hierarchy) or the lightest (inverted mass hierarchy). The determination of the mass hierarchy will constitute a large step towards the full knowledge of the PMNS neutrino mixing matrix and its potentially non-zero CP-violating phase. It is also an important input for the interpretation of $0\nu2\beta$-decay experiments.

2. The IceCube detector

IceCube [5] is a cubic-kilometre neutrino detector at the South Pole, consisting of 5160 optical light sensors at depths between $\sim$ 1.5 and $\sim$ 2.5 km below the surface which detect the light generated by charged particles produced in ν CC and NC interactions. The majority of the optical modules are deployed with a vertical spacing of 17 m and a horizontal spacing of 125 m. To increase the sensitivity in the energy range from $\sim$ 10 GeV to $\sim$ 100 GeV, a region of denser instrumentation called DeepCore [6] is located in the lower part of the detector center.

Among other detection channels, IceCube and DeepCore record atmospheric ν_μ CC interactions at neutrino energies E between $\sim$ 10 GeV and $\sim$ 100 TeV. Outgoing muon tracks with lengths of $\mathcal{O}$(50m) and more allow to distinguish these events from other interactions, and facilitate the reconstruction of the neutrino direction and thus the propagation length L. Thanks to the large field-of-view and energy range of IceCube, a large range of values of L/E can be probed. Muon neutrino disappearance has been observed in energy and energy vs zenith distributions using data from the complete and almost complete IceCube detector [7, 8, 9].

[1] http://icecube.wisc.edu/collaboration/authors/pingu

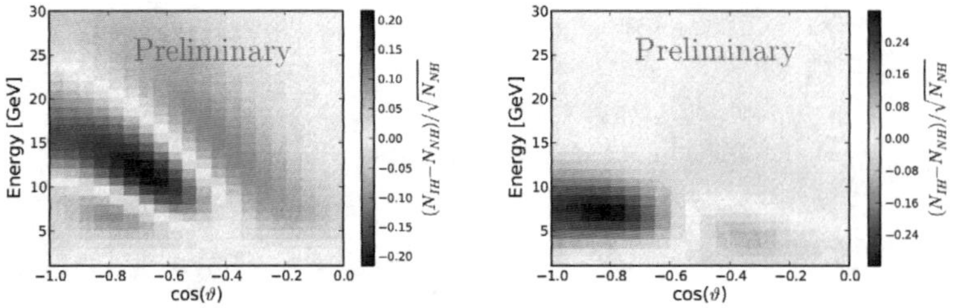

Figure 1. Illustration of impact of the neutrino mass hierarchy on the distribution of track-like (left) and cascade-like (right) events in PINGU, for 1 year of data with 2400 PINGU optical modules. The figures include the effects of reconstruction and particle classification. From [10].

3. PINGU as an extension of IceCube

PINGU is a proposed extension of the IceCube detector. The installation of additional optical modules in the deep center of IceCube with vertical spacing of $\mathcal{O}$(5m) and horizontal spacing of $\mathcal{O}$(20m) will lower the energy threshold to a few GeV. The location inside IceCube will allow to use the existing array as an active veto against the ~ 6 orders of magnitude larger background of down-going atmospheric muons. Methods to efficiently reject muon events based on the light detected in the outer parts of the detector exist and achieve a background rejection of 10^7 [11].

The new sensors can be deployed within a very short time frame of 2-3 years; for IceCube up to 1200 optical modules were installed at 20 different horizontal positions in a single Antarctic summer. The full integration of PINGU into IceCube's on-line and off-line software systems implies minimal cost for software development, a seamless combination of the data and a fast turnaround in the analysis of the data. The design studies for PINGU use the simulation framework of IceCube and are thus able to fully include the optical properties of the ice, the data acquisition, the topology of the expected events and the most important systematic uncertainties.

PINGU may also provide a positive feedback to IceCube. The design and analysis of data from DeepCore have for example resulted in advanced noise cleaning algorithms [12] and extensions of the simulation. The first evaluation of an atmospheric neutrino veto [14], a result that is now also used at very high energies [13], was inspired by DeepCore. PINGU will also provide the opportunity to install new calibration devices in the center of IceCube.

4. PINGU and the neutrino mass hierarchy

As atmospheric neutrinos travel through the Earth, their oscillation patterns are changed by matter effects which depend on the NMH. The Mikhaev-Smirnov-Wolfenstein effect [15, 16] and parametric resonances [17] at the core-mantle boundary enhance the $\nu_\mu \to \nu_e$ ($\bar{\nu}_\mu \to \bar{\nu}_e$) probability at certain L/E in the normal (inverted) hierarchy. PINGU will measure the sum of ν and $\bar{\nu}$ interactions as a function of zenith angle and energy distinguishing between events with and without evidence of a track (track-like vs cascade-like). Due to differences in the atmospheric flux and the cross-sections of ν and $\bar{\nu}$, there will be an imprint of the NMH in the distribution of the detected events. Figure 1 (from [10]) shows the difference in the expected number of events under the normal (N_{NH}) and inverted hierachy (N_{IH}) after one year, normalized to $\sqrt{N_{NH}}$. 2400 additional optical modules at 40 horizontal positions with a vertical spacing of 5 m were simulated. A reconstruction of arrival direction, energy and vertex position was applied and the vertex was required to be within or below the instrumented volume of PINGU. A parametrization of PINGU's expected capability to distinguish track- and cascade-like events was used. The left plot shows track-like events (mostly ν_μ and $\bar{\nu}_\mu$ CC events) while the right plot shows cascade-like

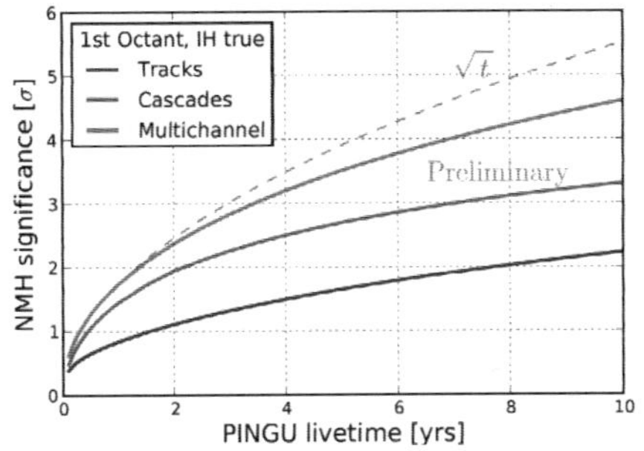

Figure 2. Expected significance for the rejection of the normal mass hierarchy as a function of time, assuming that the inverted hierarchy is true and including the main systematic uncertainties with a parametric approach using the same detector configuration as described above. Higher significances are expected for $\Theta_{23} > 45°$. From [10].

events. The NMH leaves a clear imprint in these distributions.

Figure 2 (from [10]) shows for the same detector configuration and event selection the estimated median significance for the rejection of the normal hierarchy as a function of time, assuming that the inverted hierarchy is realized in nature and that Θ_{23} is in the first octant. A full analysis of the PINGU data will be a likelihood analysis of the two-dimensional event pattern including systematic uncertainties as nuisance parameters. The sensitivity shown here has been obtained with a faster approach based on the Fisher matrix [18]. It has been shown that this approach yields similar although slightly lower sensitivities than the full likelihood analysis [10]. The estimated significance takes an energy scale uncertainty and an energy-dependent uncertainty of the effective area into account. Also a 15% uncertainty on the $\nu/\bar{\nu}$ cross-sections and uncertainties on the oscillation parameters Δm^2_{31}, Θ_{31} and Θ_{23} are included.

PINGU's configuration can be optimized for the measurement of the NMH. Advanced particle identification and reconstruction methods may in the future further improve the sensitivity, as well as improved measurements of the neutrino cross-sections and oscillation parameters. Beyond the NMH measurement, PINGU will also measure the atmospheric oscillation parameters at high precision and improve IceCube's sensitivity to dark matter and supernovae [10].

References

[1] Y Fukuda et al. [Super-Kamiokande Collaboration], Phys. Rev. Lett. **81** (1998) 1562.
[2] Q R Ahmad et al. [SNO Collaboration], Phys. Rev. Lett. 87 (2001) 071301.
[3] P Adamson et al. [MINOS collaboration], Phys. Rev. Lett. **110** (2013) 251801.
[4] F P An et al. [Daya Bay collaboration], Phys. Rev. Lett. **108** (2012) 171803.
[5] A Achterberg et al. [IceCube collaboration], Astropart. Phys. **26** (2006) 155-173.
[6] R Abbasi et al. [IceCube collaboration], Astropart. Phys. **35** (2012) 615-624.
[7] M G Aartsen et al. [IceCube collaboration] Phys. Rev. Lett. **111** (2013) 081801.
[8] S Euler, L Gladstone, C Wiebusch for the IceCube collaboration, *Measurement of atmospheric neutrino oscillations with IceCube/DeepCore in its 79-string configuration*, Proceedings of the 33rd International Cosmic Ray Conference (2013) , Rio de Janeiro, Brazil., arXiv:1304.0735.
[9] J P Yáñez for the IceCube collaboration, *Measurement of Neutrino Oscillations with the Full IceCube Detector*, Proceedings of the 33rd International Cosmic Ray Conference (2013) , Rio de Janeiro, Brazil., arXiv:1304.0735.
[10] M G Aartsen et al. [IceCube-PINGU collaboration], *Letter of Intent: The Precision IceCube Next Generation Upgrade*, arxiv:1401.2046 (2013).
[11] M G Aartsen et al. [IceCube collaboration], Phys. Rev. Lett. **110** (2013) 151105.
[12] O Schulz, *The design study of IceCube-DeepCore: Characterization and veto studies*, PhD Thesis, Universität Heidelberg (2010).
[13] M G Aartsen et al. [IceCube collaboration] Science **342**, 1242856 (2013).
[14] S Schoenert, T K Gaisser, E Resconi, O Schulz, PRD **79** (2009) 043009.
[15] L Wolfenstein, Phys. Rev. **D17** (1978) 2369-2374.
[16] S Mikheyev and A Y Smirnov, Prog. Part. Nucl. Phys. **23** (1989) 41-136.
[17] E K Akhmedov, A Dighe, P Lipari, and A Y Smirnov, Nucl. Phys. **B542** (1999) 3-30.
[18] R A Fisher, Phil. Trans. Royal Soc. A **220** (1922) 309-368.

Neutrino parameters and the N_2-dominated scenario of leptogenesis

M Re Fiorentin[1] and SE King[1,2]

[1] School of Physics and Astronomy, University of Southampton, Southampton, SO17 1BJ, UK
[2] School of Physics and Astronomy, Queen Mary University of London, London, E1 4NS, UK

E-mail: m.re-fiorentin@soton.ac.uk, soph.e.king123@gmail.com

Abstract. We briefly review the main aspects of leptogenesis, describing both the unflavoured and the flavoured versions of the N_2-dominated scenario. A study of the success rates of both classes of models has been carried out. We comment on these results and discuss corrective effects to this simplest scenario. Focusing on the flavoured case, we consider the conditions required by strong thermal leptogenesis, where the final asymmetry is fully independent of the initial conditions. Barring strong cancellations in the seesaw formula and in the flavoured decay parameters, we show that strong thermal leptogenesis favours a lightest neutrino mass $m_1 \gtrsim 10$ meV for normal ordering (NO) and $m_1 \gtrsim 3$ meV for inverted ordering (IO). Finally, we briefly comment on the power of absolute neutrino mass scale experiments to either support or severely corner strong thermal leptogenesis.

1. Leptogenesis and the N_2-dominated scenario

Leptogenesis is a particularly attractive process for producing the baryon asymmetry of the Universe, since it utilises the same mechanism that is able to explain the observed neutrino masses and mixing, namely the seesaw mechanism. Considering type-I seesaw, the Standard Model is extended by including heavy right-handed (RH) Majorana neutrinos, N_i, that couple to the lepton doublets via Yukawa interactions. Their CP- and lepton-number-violating decay produces a lepton asymmetry, which is partly converted into the baryon sector by means of sphaleron processes. We assume in general the presence of 3 RH neutrinos, thus giving rise to a model that depends on 18 parameters. These split into the 9 so-called *low-energy neutrino parameters,* probed in experiments, and a further 9 describing the high-energy scale of the RH neutrinos. We relate these two sets of parameters via the complex orthogonal matrix Ω parameterisation [1]. By definition, N_2*-dominated* [2] scenarios produce the observed asymmetry by the decay of the next-to-lightest RH neutrino, N_2. The subsequent decay of the lightest RH neutrino, N_1, produces a negligible contribution and, furthermore, acts to wash the asymmetry out. These models are particularly interesting because they naturally arise when conditions inspired by $SO(10)$ grand unification models are considered. Indeed, in this framework the spectrum of the RH neutrinos is hierarchical with $M_3 > 10^{12}$ GeV, 10^9 GeV $< M_2 < 10^{12}$ GeV and $M_1 < 10^9$ GeV, hence leptogenesis is N_2-dominated. In the unflavoured case, flavour interactions are neglected and RH neutrinos decay into a coherent superposition of flavour eigenstates. The final value of the asymmetry is exponentially suppressed by the unflavoured N_1 decay parameter, $K_1 \equiv \Gamma_1/H(T = M_1)$, where H is the Hubble parameter and Γ_1 is the N_1

total decay width. Only when $K_1 \ll 1$ is the model able to produce the observed asymmetry. A numerical analysis shows that this is realised in only $\sim 0.2\%$ of the parameter space.

This picture improves by taking account of flavour interactions. Considering that τ-interactions are efficient for $T \lesssim 10^{12}$ GeV, and muonic ones for $T \lesssim 10^9$ GeV, N_2 decays in a two fully-flavoured regime 10^9 GeV $\lesssim T \lesssim 10^{12}$ GeV, while the wash-out due to N_1 takes place in a three fully-flavoured regime $T \lesssim 10^9$ GeV. Around the transition temperatures the behaviour cannot be described by the usual Boltzmann equations and a density matrix formalism must be adopted [3]. Avoiding these regions, the asymmetry produced by N_2 is projected onto the flavour basis and the N_1 wash-out acts separately on each flavour α ($\alpha = e, \mu, \tau$), exponentially suppressing the asymmetry by the relative flavoured decay parameter, $K_{1\alpha} \equiv (\Gamma_{i\alpha} + \bar{\Gamma}_{i\alpha})/H(T = M_1)$. This only requires one $K_{1\alpha} \ll 1$ in order to have a successful model. This is found to happen in $\sim 30\%$ of the parameter space, showing that thanks to flavour effects the N_2-dominated scenario can represent a viable model of leptogenesis.

Some approximations are made: in particular, the presence of phantom terms and flavour coupling is neglected [4]. The phantom terms correct the projection onto the flavour basis of the asymmetry produced by N_2 in the coherent superposition of e and μ flavours. Each flavour asymmetry can be regarded as a combination of a term proportional to the total lepton asymmetry and a term related to the different flavour composition of the lepton quantum states with respect to their CP conjugate. This contribution is commonly referred to as a *phantom term*. A further effect, *flavour coupling*, introduces a correction when one considers that the Boltzmann equations are coupled due to the asymmetries stored in the lepton doublets and in the Higgs bosons. While the size of these effects has already been studied, we focus our attention on the size of the parameter space region where these corrections become significant. A numerical analysis reveals that for $\sim 20\%$ of the parameter space the phantom terms and the flavour coupling modify the final asymmetry by at least one order of magnitude.

2. Strong thermal leptogenesis

The final baryon asymmetry depends in general on the initial conditions. At the high temperatures required by this minimal scenario, it is possible that prior to leptogenesis some other mechanisms may generate a significant asymmetry. Leptogenesis models able to wash out a pre-existing asymmetry $N^{p,i}_{B-L}$, while producing the correct final amount, are said to satisfy the *strong thermal leptogenesis* condition [5]. It has been shown [5] that this case is realised only in a two-stage process where the RH neutrinos adhere to the pattern discussed above, thus reinforcing the interest in N_2-dominated models. In this scenario, N_2 decays in the two-flavoured regime and efficiently washes out the τ component of $N^{p,i}_{B-L}$. N_1, on the other hand, decays in the three-flavoured regime, separately erasing the e and μ components of $N^{p,i}_{B-L}$, while the produced asymmetry survives in the τ flavour. This translates into a set of conditions on the relevant decay parameters, namely $K_{1e}, K_{1\mu}, K_{2\tau} \gg 1$ and $K_{1\tau} \lesssim 1$. When the low-energy neutrino parameters are constrained within the current experimental ranges, our analysis shows that, in the NO case, the strong thermal conditions can be simultaneously satisfied, without fine-tuning, only if the absolute neutrino mass scale m_1 is sufficiently large. In particular, for small values of m_1 it becomes difficult to obtain a value of K_{1e} large enough to ensure the wash-out of a pre-existing asymmetry, while keeping $K_{1\tau}$ small. It is then possible to place a precise analytical lower bound on m_1 [6]. This lower bound depends on the magnitude of $N^{p,i}_{B-L}$ and on the size of the entries of the orthogonal matrix Ω, i.e. on $\max[|\Omega_{ij}|^2] \equiv M_\Omega$. The larger M_Ω, the more the seesaw mechanism relies, in order to work, on fine-tuned cancellations, rather than on the natural interplay between the electroweak scale and the higher, beyond the Standard Model, scale of the RH neutrinos. In general, the lower bound shows a characteristic dependence on the Dirac phase, δ, and has its minimum for $\delta = 0$. For $M_\Omega \leq 2$ and a large pre-existing asymmetry, $N^{p,i}_{B-L} = 0.1$, at $\delta = 0$ one gets $m_1 \geq 0.7$ meV at 95% C.L. The lower bound is relaxed for lower values of $N^{p,i}_{B-L}$ and larger values of M_Ω. In particular, for $M_\Omega \gtrsim 4$ the lower bound disappears. Our results are fully supported by a numerical analysis that demonstrates the analytical lower bound being saturated at the expense of a high level of fine-tuning in the seesaw formula and in the flavoured decay parameters. Moreover, a statistical study of the m_1 distribution shows a clear peak around $m_1 \sim m_{\rm atm} \simeq 50$ meV and rapid suppression for smaller values. In the aforementioned setup, more than 99% of points are found for $m_1 \gtrsim 10$ meV (the value quoted in the abstract).

In the IO case, tension exists between the smallness of $K_{1\tau}$ and a $K_{1\mu}$ that must be large enough to wash-out the pre-existing asymmetry. However, with respect to the NO case, $K_{1\mu}$ is easily large enough, even for small values of m_1. More precisely, an analytical lower bound on m_1 holds for IO only when $M_\Omega \lesssim 0.9$ and is therefore much looser than for NO. The numerical analysis confirms again the analytical results, but now 99% of the points are found for $m_1 \gtrsim 3$ meV (the value quoted in the abstract).

It is important to stress again that these results rely on the need for a K_{1e} ($K_{1\mu}$), in NO (IO), large enough to efficiently wash-out the pre-existing asymmetry, while keeping $K_{1\tau}$ small. This request becomes nontrivial due to the experimental values of the neutrino mixing angles, in particular to the suppression introduced by $\sin\theta_{13}$. It can be shown that, with our usual requirements $N^{p,i}_{B-L} = 0.1$ and $M_\Omega \leq 2$, there is no lower bound on m_1 for $30^\circ \lesssim \theta_{13} \lesssim 50^\circ$. The reactor mixing angle happens to be large enough to allow both itself and the Dirac phase to be probed, but not too large to trivially satisfy strong thermal leptogenesis and remove any lower bound on m_1.

In the coming years, experimental results of the neutrino mass measurements, especially from cosmology, may have the potential to test models of leptogenesis. A measurement of the sum of the neutrino masses can put a constraint on m_1 and, if the uncertainty is further reduced to ~ 10 meV [7], it will become possible to discriminate between vanishing and non-vanishing m_1. Hence, future cosmological observations have the power to either support or severely corner strong thermal leptogenesis. However, this is highly dependent on the ordering and therefore it is of the utmost importance that, in the next years, neutrino oscillation experiments solve the ambiguity between NO and IO. The former is in general a much more favourable case than IO for a significant test, since it hints at larger deviations from the hierarchical limit ($m_1 = 0$).

3. Conclusions

Strong thermal leptogenesis introduces a theoretical request that is able to constrain the parameter space and provide interesting results on m_1. In a reasonable setup, favourable values of the absolute neutrino mass scale are found for NO, $m_1 \gtrsim 10$ meV and IO, $m_1 \gtrsim 3$ meV. In NO a further analytical lower bound is found for natural choices of Ω ($M_\Omega \lesssim 4$), while in IO the analytical threshold is obtained only for $M_\Omega \lesssim 0.9$. These constraints on m_1 allow the future cosmological observations, together with the determination of the neutrino mass ordering, to test strong thermal leptogenesis. We conclude by highlighting that the link between cosmology and neutrino phenomenology can prove to be once again extremely fertile and rich of ideas that may shed more and more light on some of the major puzzles of modern physics.

References

[1] Casas J A, Ibarra A, 2001 *Nucl. Phys.* B **618** 171.
[2] Di Bari P, 2005 *Nucl. Phys.* B **727** 318-354.
[3] Blanchet S, Di Bari P, Jones D A, Marzola L, 2013 *JCAP* **1301** 041
[4] Antusch S, Di Bari P, Jones D A, King S F, 2012 *Nucl. Phys.* B **856** 180
[5] Bertuzzo E, Di Bari P, Marzola L, 2011 *Nucl. Phys.* B **849** 521-548.
[6] Di Bari P, King S E, Re Fiorentin M, 2014 *JCAP* **1403** 050
[7] Hamann J, Hannestad S, Wong Y, 2012 *JCAP* **1211** 052

Decays in Flight of Heavy Sterile Neutrinos in the MiniBooNE Detector

Mark Ross-Lonergan[1]

1 Institute of Particle Physics Phenomenology, University of Durham, Durham UK

E-mail: mark.ross-lonergan@durham.ac.uk

Abstract. Heavy sterile neutrinos in the $\mathcal{O}(10)$ MeV mass range would be produced in neutrino superbeams, from pion and kaon decays. In short-baseline experiments and for sufficiently long lifetimes, such steriles can propagate to the detector and decay to electrons or photons, leading to signatures which may be mis-identified for oscillation driven electron quasi-elastic charged current events. Some generic features of the production and decay of such steriles are mentioned, and the particular case study of the low energy excess at the MiniBooNE detector is studied for possible signals. We show that angular data heavily constraints the amount of events that could be from such decays to at most 50 % of the excess.

The MiniBooNE experiment was proposed and built due to the anomalous results obtained during the running of the LSND experiment[1], which could be interpreted as neutrino oscillations with a mass difference of approximately $\Delta m^2 \approx 1$ eV^2, requiring an additional *sterile* neutrino when combined with the limits on active neutrinos from Z decay precision measurements[2]. Although no conclusive evidence exists, the MiniBooNE data alongside the previous results from LSND and reactor anomalies [3] represent a hint of new physics at the eV scale. The excess low-energy (< 400 MeV) events observed in MiniBooNE , however, do not agree well with standard 3+1 or 3+2 explanations for the anomalies.[4]

As such, alternative explanations are sought for the low-energy excess in isolation, and in these proceedings the possibility of a signal being generated by a MeV scale sterile neutrino decaying in flight is investigated. Although no evidence currently exists to suggest new particles at such an energy scale, MeV sterile neutrinos would naturally be produced at neutrino beam facilities around the world through standard meson decay, with an extra factor of the appropriate mixing element suppressing the production, allowing for a rich phenomenology in many present and future experiments.

A common method in the literature to estimate the expected flux of such sterile neutrinos contained in a standard neutrino beam, is by modifying the known neutrino flux by the mixing element $|U_{\mu 4}|^2$. This method does not lend itself to the study of any spectral information as the difference in the kinematical production of a massless and MeV state can be significant, as described by Asaka et al [5]. Other effects such as the removal of helicity suppression in the $\pi^+ \to e^+\nu_e$ channel if the electron neutrino is replaced by a MeV sterile $\pi^+ \to e^+\nu_s$, can greatly enhance fluxes and must be taken into account.

Journal of Physics: Conference Series **598** (2015) 012028 doi:10.1088/1742-6596/598/1/012028

Sterile Production at MiniBooNE

In this work we investigate the behaviour of such MeV range sterile neutrinos in modern short-baseline experiments, and for the purpose of this proceedings, we use as a case study the possibility that the MiniBooNE low-energy excess[6] is the result of decaying heavy steriles in flight.

In order to more accurately account for the kinematics of the production mechanism we follow the methods laid out by Asaka et al. in which the steriles fluxes are calculated by the integrating the parent meson spectrum (e.g Pions) over all kinematical variables such that the resultant sterile successfully impinges on the MiniBooNE detector approximately $R = 550$m downstream of the proton beam target. This includes the effects of a line source of decays, from a parent flux ϕ_π, as estimated from the HARP and E910 experiments [7].

$$\phi_s(E_s) \propto \int_0^{50m} dl \int_{\cos\theta^-}^{1} d\cos\theta \int_0^\infty dp_\pi \phi_\pi(p_\pi) \exp\left(\frac{-l\Gamma_\pi m_\pi}{p_\pi}\right) \frac{d^2\Gamma(\pi \to \nu_s \mu)}{dE_s d\cos\theta} \left(\frac{m_\pi}{p_\pi}\right), \quad (1)$$

with $\cos\theta^-$ forming a geometric acceptance cone given by the point of pion decay l and the radius of the MiniBooNE detector r , $\cos\theta^- = \cos\left(\frac{r}{R-l}\right)$

Although the total decay width $\Gamma(\pi \to \mu\nu_s)$ indeed approaches 0 as the sterile's mass approaches the kinematic boundary $m_\pi - m_\mu$, the doubly differential decay width $\frac{\partial^2\Gamma}{\partial E_s \partial\cos\theta}$ at values such that the angle is within the acceptance cone, *increases* until a turnover very close the boundary ($\ll$1 MeV from boundary). This is easily understood from within the pion rest frame; as one approaches the kinematic boundary the sterile is being produced with vanishing velocity. When then boosted back into the lab frame, this corresponds to a lab velocity increasingly parallel to the parent pion. As the pions are focused to the detector this means that there is little spread outside $\cos\theta^-$ when compared to lighter neutrino behaviour. Such a boost further increases the estimated sterile flux at the detector over that naively expected from a $|U_{\mu 4}|^2$ scaling.

Sterile Decay

Once produced, and if the lifetime is sufficiently large, the steriles propagate the baseline length and their decays can be observed inside the detector. A decay such as the radiative $\nu_s \to \nu_f \gamma$ channel gives rise to a individual photon inside the detector, whose EM shower is indistinguishable from that of an electrons, in the water based cherenkov detector. As such it is mid-identified as an electron from an oscillated quasi-elastic charged current event. To keep this as model independent as possible the exact method of decay is keep arbitrary and is quantified by an effective decay width Γ_{eff}. The flux of observed photons for a given visible energy E_γ and total sterile decay width Γ_{tot} is

$$\phi_\gamma(E_\gamma) = \Gamma_{\text{eff}} \int_{E_0}^\infty \phi_s(E_s) \exp\left(\frac{-(R-l)\Gamma_{\text{Tot}} m_s}{p_s}\right) \frac{m_s}{p_s^2}, \quad (2)$$

with E_0 being the kinematic threshold $E_0 = E_\gamma(1 - m_s^2/4E_\gamma^2)$.

Energy and Angular Spectrums at MiniBooNE

Quantitatively, from a fit to the energy spectrum only, the significance for rejecting the background only model in favour of that of a decaying sterile is approximately 3σ with systematic errors in the 5-10% level, if the sterile is from a parent pion. Steriles with a kaon origin, however, never exceed 2σ due to the heavier parent smearing the spectrum to higher energies.

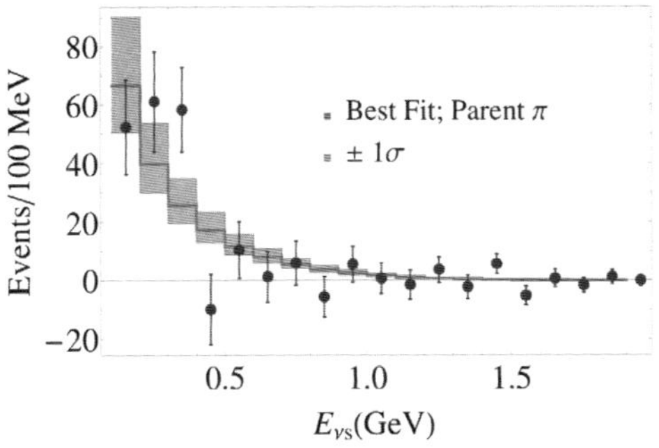

Figure 1: Energy spectrum for a 25 MeV sterile, with $|U_{\mu 4}|^2\Gamma_{\rm eff} = 10^{-29.4}$ GeV in comparison to MiniBooNE low-energy excess

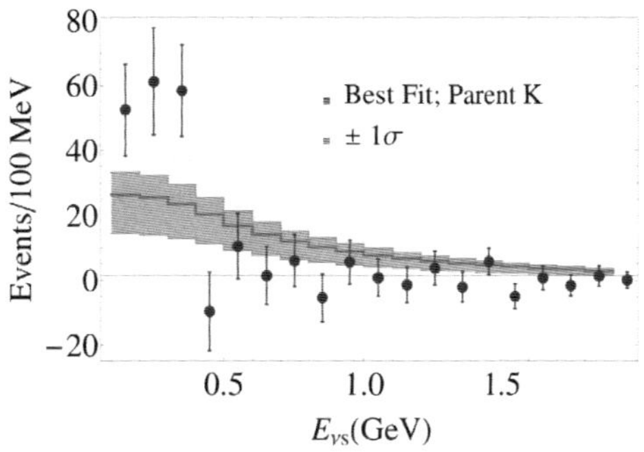

Figure 2: Energy spectrum for a 300 MeV sterile, with $|U_{\mu 4}|^2\Gamma_{\rm eff} = 10^{-30.51}$ GeV in comparison to MiniBooNE low-energy excess.

MiniBooNE does not just report on the energy spectrum, however, is also has angular information for its events. The decay in flight mechanism produces decay products which are naturally boosted forward significantly. Hence the angular spectrum generated by such processes is a much better variable to constrain the amount that such decays can contribute to the observed low-energy excess. The heavier steriles from parent kaons actually fit the angular spectrum better due to possibility of much heavier daughter particles not being boosted to such a degree.

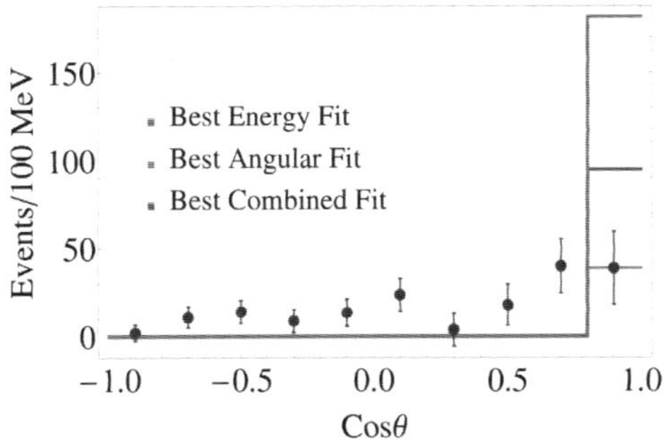

Figure 3: Angular spectrum for the above 25 MeV sterile, showing the disagreement due to large forward boosting of daughter photon

A conservative estimate using angular data is that, at the 90% C.L, the fraction of excess events that can originate from such sterile decays at $\approx$ 50%, with the significance of rejecting a background only model over such a scenario *alone* dropping to $\approx 1\sigma$. As such we conclude that the simplest model of single photons from heavy sterile decays cannot adequately explain the low-energy excess, due to the interplaying kinematics that mean a better energy fit requires regions of parameter space that directly leads to a worse angular spectrum, and vice versa.

Although the simplest model fails to reproduce the MiniBooNE excess, it highlights the phenomenology of such decays in modern experiments, and also raises the possibility that one could use such decays to place bounds on sterile-active mixing combinations such as $|U_{\tau 4}||U_{\mu 4}|$, which currently suffers from lack on knowledge on $U_{\tau 4}$. Furthermore, by considering more complicated decays, as well as the possibility of decays *before* the detector one may be able to modify the final state kinematics so that a more isotropic spectrum is generated, easing tension between the MiniBooNE angular and energy data. Further work is undergoing.

Acknowledgments
These proceedings are based on work with Silvia Pascoli and Peter Ballett. This work is funded by Invisibles, a European ITN project (FP7-PEOPLE-2011 -ITN, PITN-GA-2011 -289442-INVISIBLES (April 2012-March 2016))

References
[1] Church E, Eitel K, Mills G B and Steidl M 2002 *Phys.Rev.* **D66** 013001 (*Preprint* hep-ex/0203023)
[2] Schael S *et al.* (ALEPH Collaboration, DELPHI Collaboration, L3 Collaboration, OPAL Collaboration, SLD Collaboration, LEP Electroweak Working Group, SLD Electroweak Group, SLD Heavy Flavour Group) 2006 *Phys.Rept.* **427** 257–454 (*Preprint* hep-ex/0509008)
[3] Mention G, Fechner M, Lasserre T, Mueller T, Lhuillier D *et al.* 2011 *Phys.Rev.* **D83** 073006 (*Preprint* 1101.2755)
[4] Giunti C, Laveder M, Li Y and Long H 2013 *Phys.Rev.* **D88** 073008 (*Preprint* 1308.5288)
[5] Asaka T, Eijima S and Watanabe A 2013 *JHEP* **1303** 125 (*Preprint* 1212.1062)
[6] Aguilar-Arevalo A *et al.* (MiniBooNE Collaboration) 2013 *Phys.Rev.Lett.* **110** 161801 (*Preprint* 1207.4809)
[7] Aguilar-Arevalo A *et al.* (MiniBooNE Collaboration) 2009 *Phys.Rev.* **D79** 072002 (*Preprint* 0806.1449)

Journal of Physics: Conference Series **598** (2015) 012029 doi:10.1088/1742-6596/598/1/012029

Cosmological bounds on active-sterile neutrino mixing after Planck data

Ninetta Saviano

Institute for Particle Physics Phenomenology, Department of Physics, Durham University, Durham DH1 3LE, UK

E-mail: ninetta.saviano@durham.ac.uk

Abstract. Light sterile neutrinos can be produced by oscillations with active neutrinos in the early universe. Their properties can be constrained by their contribution as extra-radiation, parametrized in terms of the effective number of neutrino species $N_{\rm eff}$, and to the universe energy density today $\Omega_\nu h^2$. Both these parameters have been measured to quite a good precision by the Planck satellite experiment. We use these results to update the bounds on the parameter space of (3+1) sterile neutrino scenarios, with an active-sterile neutrino mass squared splitting in the range $(10^{-5} - 10^2)\,{\rm eV}^2$. For the first time we take into account the possibility of two non-vanishing active-sterile mixing angles. We find that the mass and mixing parameter space is severely constrained. In particular sterile neutrinos with $m \sim \mathcal{O}(1)$ eV are strongly disfavoured by neutrino mass bound.

1. Introduction

In recent years a renewed attention has been devoted to low-mass sterile neutrinos ($m \sim$1 eV), after intriguing but controversial hints coming from laboratory oscillation experiments. Many analyses have been performed to explain the anomalies in scenarios with one (dubbed "3+1") or two ("3+2") eV sterile neutrinos [1, 2]. In this context, cosmological observations represent an important tool, being sensitive to the number of neutrinos and to their mass at eV scale. Indeed, sterile neutrinos can be produced by oscillations with the active ones in the early universe participating, if sufficiently light, to the radiation content, parametrized in terms of $N_{\rm eff}$, leaving possible traces on different cosmological observables. Moreover they can contribute to the energy density in the universe today. The Big Bang Nucleosynthesis (BBN) marginally allows one fully thermalized sterile neutrino. The Cosmic Microwave Background (CMB) measurements have shown a preference for extra radiation which has been reduced over the years. Moreover, even a single extra thermalized sterile neutrino with mass $m \sim 1$ eV appears to be inconsistent with mass bounds from CMB and LSS data. A recent and important contribution in determining the *dark* radiation content in the early universe and in constraining the neutrino mass bound is represented by the first data release of the Planck experiment providing, in combination with WMAP, Baryon Acoustic Oscillation and high multipole CMB data, $N_{\rm eff} = 3.30 \pm 0.27$ at 68 % C.L. [3]. Given this partially contradictory situation it is necessary to assess the sterile abundance through the flavor evolution of the active-sterile neutrino ensemble. In particular, motivated by the recent data release of the Planck experiment, we performed an extensive scan of the sterile neutrino parameter space in a 3+1 model, with the sterile mass splitting $\Delta m^2_{\rm st}$ in the range $(10^{-5} - 10^{2})\,\mathrm{eV}^2$ and considering for the first time the possibility of two non-vanishing active-sterile mixing angles. Once more we confirm the tension between the cosmological data and the mass and mixing suggested by the laboratory hints.

2. Setup of the (3+1) flavor evolution

We consider one single light sterile neutrino ν_s, which mixes with the active neutrino states ν_e, ν_μ, ν_τ. In our study we fix the values of the three active mixing angles to the current best-fit from global analysis of the different active neutrino oscillation data, $\sin^2\theta_{12} = 0.307$, $\sin^2\theta_{23} = 0.398$, and $\sin^2\theta_{13} = 0.0245$. Concerning the active-sterile mixing angles we choose as representative range $10^{-5} \leq \sin^2\theta_{i4} \leq 10^{-1}$ (i=1,2,3).

The solar and the atmospheric mass-square differences are given by $\Delta m^2_{21} = m^2_2 - m^2_1 = 7.54 \times 10^{-5}\ \mathrm{eV}^2$ and $|\Delta m^2_{31}| = |m^2_3 - m^2_1| = 2.43 \times 10^{-3}\ \mathrm{eV}^2$, respectively. Concerning the sterile mass-square difference, we consider the following range $10^{-5} \leq \Delta m^2_{41}/\mathrm{eV}^2 \leq 10^2$.

The neutrino (antineutrino) ensemble in the early universe plasma, is described in terms of a 4×4 momentum-averaged density matrix ρ ($\bar{\rho}$). The evolution equation for neutrino ensemble is given by:

$$\mathrm{i}\,\frac{d\rho}{dt} = [\Omega_{\mathrm{vac}}, \rho] + [\Omega_{\mathrm{mat}}, \rho] + [\Omega_{\nu-\nu}, \rho] + C[\rho]\,, \tag{1}$$

and a similar expression holds for the antineutrino matrix $\bar{\rho}$. The first term on the right-hand-side of Eq. (1) represents the vacuum term, the second and the third ones describe refractive interactions, in particular we have the ordinary matter term and the neutrino-neutrino one, respectively. Remarkably, the matter term can induce Mikheyeev-Smirnov-Wolfenstein (MSW)-like resonances when it becomes of the same order of the neutrino mass-squared splitting. In the sterile sector resonances are associated with the three different active-sterile mass splittings Δm^2_{4i} and with the different θ_{i4} mixing angles. Finally, the last term in the right-hand-side of Eq. (1) is the collisional term.

3. Cosmological bounds: results

In our investigation we use as benchmark the Planck joint constraints for extra radiation and mass. In particular, the first quantity we are exploiting to constrain the sterile neutrino parameter space is the overall non electromagnetic radiation content, parametrized via N_{eff},

$$N_{\mathrm{eff}} = \frac{1}{2}\mathrm{Tr}(\rho + \bar{\rho}) \ . \tag{2}$$

Our bounds are given comparing this number with the one measured by Planck experiment, $N_{\mathrm{eff}} < 3.80$ at 95 % C.L. [3]. Since neutrinos with mass larger then 1 eV would be not relativistic anymore at the CMB decoupling, we cannot use radiation constraints. However, mass bounds become very important through the neutrino contribution to the energy density in the Universe. Assuming the existence of a massive sterile neutrino together with two massless active neutrinos and one very light (i.e. m 0.06 eV), the second quantity is

$$\Omega_\nu h^2 = \frac{1}{2}\frac{\sqrt{\Delta m^2_{41}} \cdot (\rho_{ss} + \bar{\rho}_{ss})}{94.1\ \mathrm{eV}} \ . \tag{3}$$

The constraint on the neutrino energy density is $\Omega_\nu h^2 \leq 0.0045$ at 95 % C.L, coming from the Planck+ BAO bound on the effective sterile mass $m_s^{\mathrm{eff}} \leq 0.42$ eV. In Fig. 1 we present our exclusion plot in the planes $(\Delta m^2_{41}, \sin^2\theta_{14})$ for different values of the other mixing angle $\sin^2\theta_{24}$. In according to the analysis for the laboratory anomalies, $\sin^2\theta_{34}$ is fixed to zero. The excluded regions from N_{eff} are those on the right or at the exterior of the black contours, while

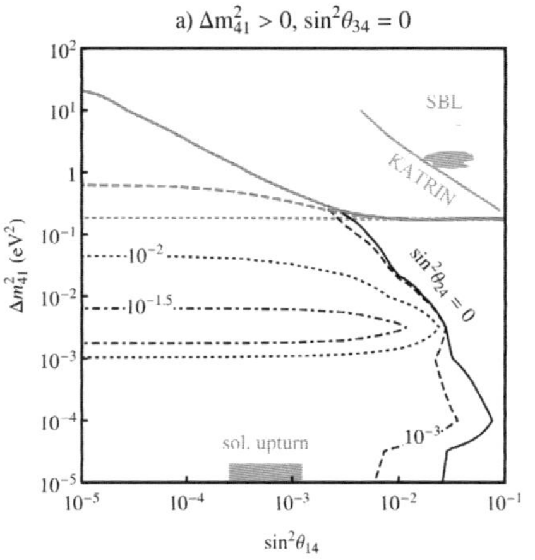

Figure 1. Exclusion plots for the active-sterile neutrino mixing parameter space from $N_{\rm eff}$ (black curves) and $\Omega_\nu h^2$ (red curves) at 95 % C.L.

the ones from $\Omega_\nu h^2$ are above the red contours. We discuss at first the bound on $N_{\rm eff}$. The most conservative limit corresponds to $\sin^2\theta_{24} = 0$, where for $\Delta m^2_{41} > 10^{-2}$ eV^2 the exclusion contour is a straight line in this plane. We cut the bound at $\Delta m^2_{41} \sim 10^{-1}$ eV^2, since for higher masses sterile neutrinos would almost be non-relativistic at the CMB epoch. Increasing the value of $\sin^2\theta_{24}$, the constraint on the parameter space becomes stronger. Large values of $\sin^2\theta_{24}$ would dominate the sterile neutrino production excluding regions otherwise permitted. Finally, we represent with a rectangle in the lower part of the plot, the region of parameters corresponding to a light sterile neutrino suggested to solve the problem of the upturn of the solar neutrino spectrum. We realize that this region is excluded if $\sin^2\theta_{24} > 10^{-3}$. We address the reader to our publication [4] for more details.

Passing now to the bounds from $\Omega_\nu h^2$, again the most conservative limit is for $\sin^2\theta_{24} = 0$. The bound becomes more stringent increasing the value of $\sin^2\theta_{24}$. In particular, for $\sin^2\theta_{24} \geq 10^{-2}$, $\Delta m^2_{41} \geq 10^{-1}$ eV^2 is excluded independently on the value of $\sin^2\theta_{14}$, since sterile neutrinos would be always produced with thermal abundances. For comparison, we also show the slice at $\sin^2\theta_{24} = 10^{-2}$ of the 95 % C.L., for the allowed region obtained from the global analysis of short-baseline oscillation data [1] (filled region in the up right part of the plot denoted by SBL). We observe that it is completely ruled out by the cosmological bound from $\Omega_\nu h^2$. A possible escape route could be represented by the suppression of the sterile neutrino production [5].

Acknowledgments

N.S. acknowledges support from the European Union FP7 ITN INVISIBLES (Marie Curie Actions, PITN- GA-2011- 289442).

References

[1] C. Giunti, M. Laveder, Y. F. Li and H. W. Long, Phys. Rev. D **88** (2013) 073008.
[2] J. Kopp, P. A. N. Machado, M. Maltoni and T. Schwetz, JHEP **1305** (2013) 050.
[3] P. A. R. Ade *et al.* [Planck Collaboration], "Planck 2013 results. XVI. Cosmological parameters," arXiv:1303.5076 [astro-ph.CO].
[4] A. Mirizzi, G. Mangano, N. Saviano, E. Borriello, C. Giunti, G. Miele and O. Pisanti, Phys. Lett. B **726** (2013) 8.
[5] N. Saviano, A. Mirizzi, O. Pisanti, P. D. Serpico, G. Mangano and G. Miele, Phys. Rev. D **87** (2013) 073006.

Journal of Physics: Conference Series **598** (2015) 012030 doi:10.1088/1742-6596/598/1/012030

Optical Calibration of SNO+.

J. Maneira[1,3], **S. Peeters**[2], **J. Sinclair**[2]

[1]Laboratório de Instrumentação e Física Experimental de Partículas, Av. Elias Garcia, 14, 1°, 1000-149 Lisboa, Portugal,

[2]Dept. of Physics and Astronomy, University of Sussex, Falmer Campus, Brighton BN1 9QH, United Kingdom,

[3] Departamento de Física, Faculdade de Ciências da Universidade de Lisboa, Campo Grande, Edifício C8, 1749-016 , Portugal

E-mail: J.Sinclair@sussex.ac.uk

Abstract. SNO is being upgraded to SNO+, which has as its main goal the search for neutrinoless double-beta decay. The upgrade is defined by filling with a novel scintillator mixture containing ^{130}Te. With a lower energy threshold than SNO, SNO+ will be sensitive to other exciting new physics. Here we are describing new optical calibration system that meets new, more stringent radiopurity requirements has been developed.

1. Introduction

SNO+ is using the infrastructure from the Sudbury Neutrino Observatory (SNO) [1]. The upgrade to SNO+ is characterised by the change in target mass: from the heavy water of SNO to a novel scintillator mixture. Situated in the SNOLAB facility, 2092 m underground in Sudbury, Northern Ontario, the depth provides SNO+ with effective shielding from cosmic muons, receiving only 3 per hour [2]. The detector consists of a 5 cm thick spherical acrylic vessel (AV) 12 m in diameter containing the target mass. The AV is mounted within a stainless steel geodesic sphere, 17.8 m in diameter. The geodesic sphere, known as the PMT support structure (PSUP), holds approximately 9500 PMTs. The PSUP and AV are suspended inside a 34 m deep cylindrical cavity, lined with Urylon as a Rn barrier. The cavity is filled with a little over 7000 tonnes of ultra pure water, providing further shielding to the inner detector from the intrinsic radioactivity of the PMTs and surrounding rocks.

Using liquid scintillator lowers the energy threshold to approximately 200 keV, making SNO+ a multifaceted experiment. SNO+'s main physics goal is addressing the question of whether neutrinos are Dirac or Majorana particles, through the search for neutrinoless double-beta decay of ^{130}Te. SNO+ will also be studying reactor, geo, supernovae and solar neutrinos, the experiment will be sensitive to ^{7}Be neutrinos during the initial commissioning phase.

The increased sensitivity to backgrounds, in particular the daughters of ^{222}Rn, imposes tight constraints on the calibration system. A number of self contained radioactive sources will be used to calibrate absolute light yield, energy linearity, geometric uniformity and alpha-beta discrimination[3]. Two complimentary systems, a deployed diffuse light source, the laserball, and the External LED/Laser Light Injection Entity (ELLIE) will be used to calibrate the

PMT timing and gain, scintillator properties and PMT response as function of event position. Source deployment must be kept to a minimum and eliminated if possible to meet radio purity requirements. This paper will provide an overview of the timing component of ELLIE, and the new laserball hardware.

2. Laserball

The laserball is a light diffusing sphere, consisting of a quartz flask filled with silicone gel, in which hollow glass spheres are suspended [4]. This diffuser is connected via optical fibres to a pulsed N_2-dye laser, which can be operated at a number of different wavelengths, allowing the scintillator to be monitored in high absorption/reemission and low absorption/scattering regimes. It was successfully used in the PMT calibration of SNO [5]. A radioactively pure laserball has been developed for SNO+, using a fully synthetic fuzed quartz flask which has less than 13 counts of ^{222}Rn per day and a neck diameter 19 mm narrower than the SNO flask, reducing shadowing to the upper PSUP. The laserball will be deployed no more than twice a year.

3. TELLIE hardware

ELLIE uses electronics mounted on a deck above the cavity, to generate light which is injected into the detector via 47.75 m long optical fibres terminating at the nodes of the PSUP. ELLIE has three sub systems: one motoring the attenuation of the scintillator, another measuring the scintillator's scattering length and the largest subsystem which provides continually calibration of PMT timing and gain (TELLIE).

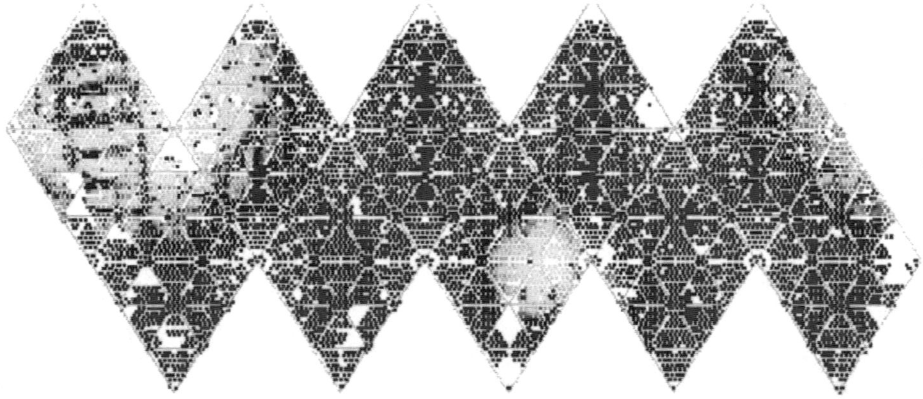

Figure 1. *A TELLIE event during the commissioning phase, the silhouette of a cleaning ladder with the AV is visible, the smaller spot is the reflection form the first surface of the AV.*

TELLIE uses 92 PMMA fibres with large opening angles ensuring overlapping PMT coverage, Figure 1 shows the coverage of a single injection point. Enough redundancy is present to bootstrap the timing calibration using overlapping regions. The fibres are connected to 12 driver boxes, each containing 8 LED drivers and PIN-diodes monitoring LED output. The drivers use transistors to flip the voltage across the LED from a reverse bias into a forward bias and then back. Low resistance LEDs improve the time profile of the pulse [6]. TELLIE has an operating range of 10^3 to 10^6 photons per pulse with widths of 4.2 ns and 7.0 ns respectively, after the full length of fibre. Figure 2 shows the a pulse of 10^3 photons, measured with a single photon counting technique.

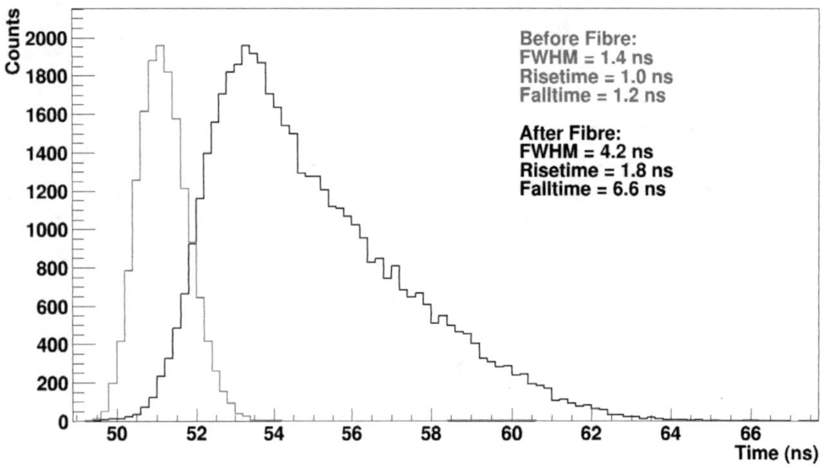

Figure 2. *A TELLIE pulse of* 10^3 *photons before and after the 47.75 m PMMA fibre, measured using a single photon counting technique.*

4. Conclusions

SNO has been upgraded to SNO+, a multipurpose neutrino detector with the primary goal of observing neutrinoless double-beta decay. Radiopurity concerns have led to the development of an external optical calibration system, which has been shown to work during air-fill, providing useful demonstration of DAQ and data flow readiness. For situations where deployment of an optical source is necessary, a much improved laserball is being developed

Acknowledgments

This research is supported by COMPETE, European Union, FCT, STFC and QREN.

References

[1] J. Boger et al., Nucl. Instr. and Meth. A 449 172-207
[2] V. Lozza, J. Phys. Conf. Series 375 042050
[3] G. Lefeuvre, Nucl. Instr. and Meth. A 718 506-508
[4] B.A. Moffat et al., Nucl. Instr. and Meth. A 544 (2005) 255-265
[5] B. Aharmim et al. Phys. Rev. C 87, 015502
[6] E. Fred Schubert, Light-Emitting Diodes, 2nd *edition*, Cambridge University Press, 2010.

Topical Research Meeting: Prospects in Neutrino Physics (NuPhys2013) IOP Publishing
Journal of Physics: Conference Series **598** (2015) 012031 doi:10.1088/1742-6596/598/1/012031

Muon reconstruction in Double Chooz

M Strait
High Energy Physics, University of Chicago, 5640 S Ellis Ave, Chicago IL 60637
E-mail: strait@hep.uchicago.edu

Abstract. We describe a muon track reconstruction algorithm for the reactor anti-neutrino experiment Double Chooz. The Double Chooz detector consists of two optically isolated volumes of liquid scintillator viewed by PMTs, and an Outer Veto above these made of crossed scintillator strips. Muons are reconstructed by their Outer Veto hit positions along with timing information from the other two detector volumes. In the ideal case of a through-going muon intersecting the center of the detector, the resolution is ~40 mm in each transverse dimension.

1. Introduction

Double Chooz is a reactor anti-neutrino experiment designed to measure the mixing parameter θ_{13} by observing inverse beta decay events, $\bar{\nu}_e p \rightarrow e^+ n$. Due to its overburden, 300 meters water equivalent, the Double Chooz far detector has a muon rate of 46 Hz. Consequently, reconstruction of muons is very helpful for suppressing cosmogenic backgrounds. The most important of these is ^{9}Li, a β-n emitter.

The design details of the detector have been described in Ref. [1]. It consists of four concentric cylindrical volumes and the Outer Veto (OV). The inner three volumes form a single optical volume isolated from the fourth and are collectively called the Inner Detector (ID). The innermost two volumes, the Neutrino Target (NT) and Gamma Catcher (GC), are filled with scintillator and are treated as a single undifferentiated volume for muon reconstruction purposes. They are surrounded by the Buffer, a volume of non-scintillating oil in a steel vessel in which 390 10-inch PMTs are placed. Outside this is the Inner Veto (IV), a volume of scintillator with 78 8-inch PMTs. Above these is the OV, a segmented plastic scintillator detector that provides (x, y) position information in 25 mm steps. It has a lower panel 1.1 m above the IV, and an upper panel 3.9 m above the lower. When a muon crosses both the upper and the lower, a high-resolution *OV track* is formed.

The characteristics of a muon event depend on which detector volumes the muon intersects. Consider muons that pass through the NT. As it traverses the ID, it first emits only Cherenkov light, then isotropic scintillation light, and then only Cherenkov again. Muons can also intersect only the IV and Buffer, or only the IV, and they can stop in any volume. In any of these cases, a muon can intersect zero, one, or two OV panels. All such combinations of through-going muons are handled by this reconstruction, as is the case of muons stopping in the ID scintillator.

2. Algorithm

Each ID and IV PMT in Double Chooz is read out using a 500 MHz flash-ADC with a readout window of 256 ns. A special pulse reconstruction is used for muons that is separate from that

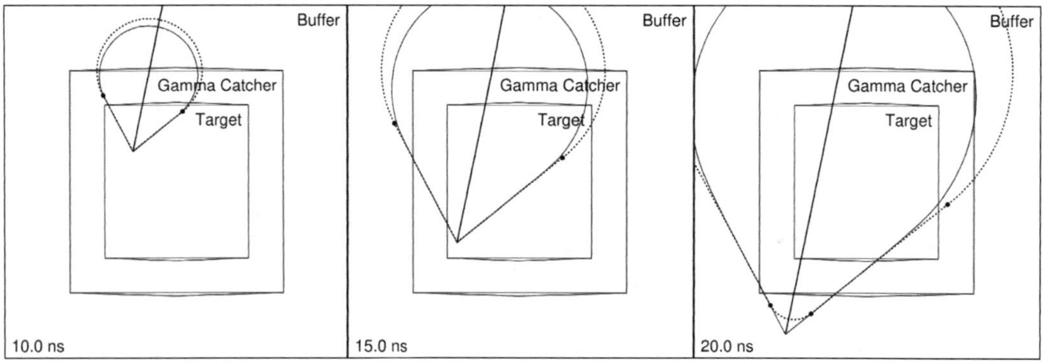

Figure 1. The wavefront of scintillator light emitted by a muon in the ID. The dashed line denotes the earliest scintillation photons. The transitions between the conical and spherical wavefronts are marked with dots. The solid line gives the wavefront used by the reconstruction.

used for other events. It defines a start time with an error, a rise time and integrated charge for each pulse. To be selected for use in the fit, IV PMTs must satisfy charge and risetime cuts to remove hits resulting from reflections. In the ID, the selection criteria are varied continuously with energy. For low-energy events, a loose selection is done that optimizes resolution for Cherenkov light. For high-energy events, a more stringent selection is used.

A χ^2 fit function is used, which MINUIT [2] minimizes. The contribution to the χ^2 from the ID is calculated either under the assumption of light mostly from Cherenkov or mostly from scintillation, depending on the event characteristics. The χ^2 is built from the selected PMTs' pulse start times and their associated errors, with unselected PMTs ignored. When OV hits are present, they are used as a spatial constraint. Since one of the goals of muon reconstruction is to look for muons with high $\mathrm{d}E/\mathrm{d}x$, we do *not* use the total reconstructed energy in the fit function for through-going muons. If the amount of light in the ID corresponds to $<30\,\mathrm{MeV}$, the code assumes a Cherenkov-dominated event. If it is $>180\,\mathrm{MeV}$, it is assumed to be scintillation-dominated. Between these, it tries both and chooses the one with the better χ^2. These criteria were tuned on data, using OV positions to check the results.

If fitting under the Cherenkov-dominated hypothesis, the PMT timings are calculated for a simple Cherenkov cone, and penalty terms are used for hits that are not intersected by the cone.

For the scintillation-dominated hypothesis, the light pattern is more complex, as shown in Fig. 1. As the muon traverses the scintillator it emits light isotropically. Due to the muon's motion, this light forms a cone similar to that of Cherenkov radiation. In addition to the cone, a sphere of light is produced behind the point where the muon enters the scintillator. The same occurs at the exit point. Cherenkov light near the entry point can generally be neglected (if significant, it is detected by the algorithm and removed), but near the exit point it is intense and overwhelms the scintillation light.

An effective shape for the wavefront is used to account for the fact that a sharp boundary between the cone and sphere shown in Fig. 1 is physically unrealistic, since it implies that an infinitesimal volume of scintillator at the entry point produces enough light to seen by every PMT that lies above the region intersected by the scintillation cone. We parametrize a corrected shape as shown, where the difference between the ideal and realistic wavefronts is controlled by a single parameter related to the muon's $\mathrm{d}E/\mathrm{d}x$. The effective speed of scintillation light is also allowed to float in the fit.

Under either fit hypothesis, IV timing information is used in the same way. Since the IV

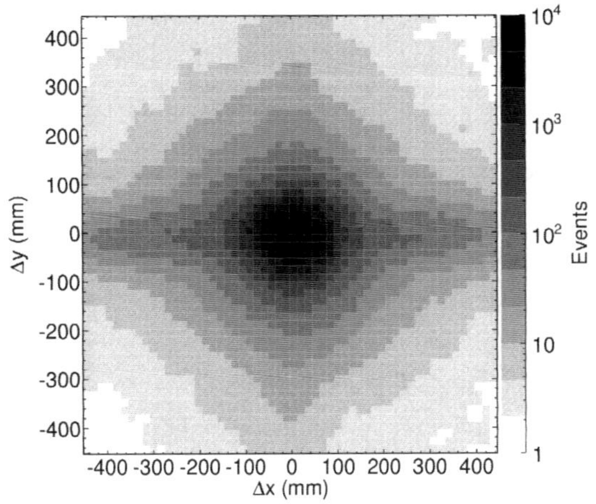

Figure 2. Resolution as measured by OV tracks. Shown is the relative position in the $z = 0$ plane between the OV track and the result of this reconstruction without use of OV data. Only events in which all ID PMTs are used are plotted here. In this case, about half of the spread is due to the OV's own resolution. The ridge along $\Delta y = 0$ arises from OV track misreconstructions and the asymmetry of the OV.

is relatively thin, an approximation is used in which the scintillation light is treated as coming from a single point for the muon's entry position and another for its exit. The expected times are calculated using these points; whichever gives a smaller χ^2 for each PMT is used. Penalty terms are added if the intersection points are not directly visible to PMTs with hits.

If the muon deposited an unusually low amount of energy in the IV, it is a stopping muon candidate. The stopping muon fit function that MINUIT minimizes is the same as the through-going scintillation-dominated fit, except that (1) light is only expected from the entry point into the IV, (2) an additional free parameter is added for the stopping point and no light is generated past this point in the model, and (3) the track energy is used to constrain the track length.

For the through-going fits, if there is OV data, the fit with the OV is usually judged to be better, but the fit without the OV is selected if it has a dramatically better χ^2. If a stopping muon fit was done, it is chosen as the best answer if the IV energy is consistent with a single IV crossing, given the reconstructed length of IV scintillator traversed.

3. Resolution

The resolution of the fit is tested using data-driven methods, primarily in events with OV tracks. We find that the resolution is about 40 mm in the ideal case of a muon intersecting a large amount of ID scintillator (see Fig. 2). For the median energy muon in Double Chooz, the limiting factor in this resolution is multiple Coulomb scattering, which violates the reconstruction's assumption of a perfectly straight track. With an (x, y) coordinates in one OV panel, the resolution is improved substantially, compared to the case of no OV data, at the top of the detector, but only marginally at the bottom.

The resolution gradually worsens as the muon's path length through the detector decreases. For shorter paths, the resolution increases by up to a factor of two from the timing-based ideal described above. For the lowest quality tracks, muons passing only through the IV, the resolution is 250 mm. Resolution for showering muons is typically 10–40% worse than for minimum ionizing muons, depending on the extent of the shower. The most important application of muon tracking in Double Chooz, ^{9}Li identification, is affected by this loss of resolution since ^{9}Li is produced by showering muons. However, the typical physical distance between the muon and the ^{9}Li production is still greater than the resolution, so the ability to identify the ^{9}Li remains excellent.

The resolution for stopping muons is 150 mm in each of x, y and z, as tested by comparing to the following Michel electron.

[1] Abe Y *et al.* (Double Chooz Collaboration) 2012 *Phys.Rev.* **D86** 052008
[2] James F and Roos M 1975 *Comput.Phys.Commun.* **10** 343–367

Journal of Physics: Conference Series **598** (2015) 012032 doi:10.1088/1742-6596/598/1/012032

Future of Neutrino Interaction Models

Ryan Terri

School of Physics and Astronomy, Queen Mary University of London, UK

E-mail: r.terri@qmul.ac.uk

Abstract. Neutrino-nucleus cross sections are one of the dominant sources of systematic errors in long-baseline neutrino oscillation experiments. To achieve the goals of precision measurements of the mixing angles and difference of the mass eigenstates squared, and discover the mass hierarchy and CP-violating phase, the underlying neutrino interactions must be better understood. This poster will mention some recent improvements in models in the interaction generators as well as some possible future improvements for proposed experiments.

1. Introduction

Long-baseline neutrino oscillation [1, 2, 3] and neutrino-nucleus scattering [4, 5] experiments have found consistent discrepancies between data and the models used in their interaction generators in the region where $E_\nu = 1\,\text{GeV}$ and between themselves. These differences contribute large systematic errors related to the neutrino cross section, and may introduce a bias in the best-fit parameters of an oscillation analysis if at least one of the interactions is being mis-modeled or is missing entirely. This paper will give a brief overview of some of the current data-model discrepancies, highlight the charged current (CC) quasi-elastic (QE) model as a source of possible improvements, and then look to the future of being able to describe describe data from current and future experiments with an eye to improving the predictions.

2. Typical Models in Neutrino Generators

A typical neutrino interaction generator, such as GENIE [6] or NEUT [7], has a base set of models to describe various interactions between a neutrino and a nuclear target. These include charged and neutral current (quasi-)elastic scattering, resonant production and decay to form single meson or photon states, coherent pion production, and deep inelastic scattering. There is also a hadron transport model used for hadrons created by an interaction in a nucleus attempting to escape, which can affect both the number and type of particles in the final state and their final outgoing momenta.

In slightly more detail, a typical model used for CCQE interactions is one developed by Llewellyn-Smith [8] in the target nucleon rest frame, with the differential cross section being:

$$\frac{d\sigma^{\nu,\overline{\nu}}}{dQ^2} = \frac{M^2 G_F^2 \cos^2\theta_C}{8\pi E_\nu^2} \times \left[A(Q^2) \mp \frac{(s-u)B(Q^2)}{M^2} + \frac{(s-u)^2 C(Q^2)}{M^4}\right], \tag{1}$$

where M is the target nucleon mass; s and u are Mandelstam variables; the -(+) sign is for (anti)neutrinos; Q^2 is non-negative; and A, B, and C are terms containing the vector form factors F_1 and F_2, pseudo vector form factor F_P, and axial form factor F_A. The axial form factor normally takes the form as a dipole

$$F_A\left(Q^2\right) = \frac{g_A}{\left(1+\frac{Q^2}{M_A^2}\right)^2}, \tag{2}$$

where g_A is a constant measured beta decay experiments and has a value of -1.267 and M_A is the axial mass. Assuming Equation 2 is the way Nature describes the axial form factor, the axial mass should have the same value in all experiments where it is measured. Values derived from pion scattering experiments and deuterium bubble chamber experiments indicate the axial mass should have a value slightly larger than 1 GeV/c^2 with only a few-% error, e.g. 1.016±0.026 GeV/c^2 [9] extracted from deuterium.

3. Discrepancies Between Data and Models

Recent cross section measurements in various neutrino experiments have highlighted some discrepancies between the interaction models and their data sets and also between experiments. For CCQE interactions, the MiniBooNE (MB) [3] and NOMAD [10] experiments measure different values of M_A, with MB reporting a central value of $M_A = 1.35$ GeV/c^2 [11] and NOMAD $M_A = 1.05$ GeV/c^2 [12].

Inclusive single pion measurements have other differences. For inclusive neutral current single π^0 measurements, the momentum spectrum for K2K [13] and MB [14] both have different shapes than the model prediction. The MB CC single pion measurements (one where the only meson is a π^+ in the final state [15], the other only a π^0 [16]) both have a shape and normalization discrepancy. Recently, MINERνA's results for CC single charged pion interactions have indicated that the default interaction models agree with their data with no significant modifications needed (this has yet to be published at the time of writing).

4. Sources of Discrepancies

The reason for these data-model discrepancies, as well as differences between experiments, is multi-fold. One source stems from the fact that most current experiments have mean neutrino energies around 1 GeV, which means lower momentum particles that have more time to interact in the nuclear environment. This implies the hadron transport models as well as the simulation of the overall nuclear density need to be treated more carefully within whatever framework is being used.

The nuclear environment also makes it so it is possible the models used in the generators are not complete. Due to the MB CCQE measurement, it has been proposed that there is a contribution from multi-nucleon interactions. Simplistically, the gauge boson is interacting with at least two nucleons, rather than one. This can be seen in models by Nieves *et al.* [17] and Martini *et al.* [18].

Finally, there is the problem of definition. Experiments and models must be clear when they use a term like CCQE due to the fact that this could mean among other things:

- $\nu_\mu + n \rightarrow \mu + p$: generator definition before intranuclear effects occur
- $\nu_\mu + X \rightarrow \mu + X' + 0$ mesons: MB's definition, where X' allows for any number of nucleons
- $\nu_\mu + X \rightarrow \mu + X' + 0$ mesons+no vertex activity: this would be a possible selection in a plastic scintillating detector
- $\nu_\mu + X \rightarrow \mu + 1p + X' + 0\pi$: NOMAD's definition

If two experiments are using the same definition, it may be easier to compare their results. Otherwise, the model needs to take these into account for cross comparisons to see if any differences possibly have the same underlying source.

5. Possible Future Improvements

The necessary improvements to the cross section models comes from the necessary dialogue between theorists, experimentalists, and phenomenologists. Since CCQE-like events (mostly interactions with a single muon and no mesons visible in the detector) have multi-nucleon contributions, being able to implement these models in a generator becomes necessary. Some steps have been taken towards this, but there needs to be an agreed-upon framework so that it becomes easy to include new models of the same phenomenon in a generator setting to test against data. Models that produce resonances are being retuned based on available data and updates to their form factors [19]. In both cases, pion-nucleon scattering measurements becomes more and more important to be able to see how we can arrive at the visible final states observed in experiment. Additionally, electron scattering data can give greater constraints on vector current form factors, helping reduce the possible parameters that can be varied in a model.

With some future very long-baseline experiments, e.g. LBNE, needing a higher neutrino energy to maximize the effect of oscillations, more attention needs to be paid to hadronization multiplicities and improving the predictions for DIS in the low-Q^2, low-W region of the interaction parameters space. Some work can be seen for improving the DIS predictions in reference [20].

While these can be done on the theory and phenomenology side of things for the generators, the community needs additional measurements made on a variety of targets across different energies. This will allow to see how the models react to data with different interactions contributing, as well as seeing how much the target affects a model's predictive power. Ideally, this would include targets from deuterium through lead, at the least, with the main focus being on carbon, oxygen, argon, and iron, since they are the most common targets used in neutrino physics. Of course, the data from these measurements also needs to be released in a way where it is possible to test any models that already exist in the generators and tune any parameters that the models allow, while also checking for consistency.

In all cases, none of these ideas are new to the community, but the hard work is making it happen on a reasonable time scale, all the while being able to react to new publications while we try to get a better handle on what Nature is telling us.

References

[1] Ahn M *et al.* (K2K Collaboration) 2006 *Phys.Rev.* **D74** 072003 (*Preprint* hep-ex/0606032)
[2] Abe K *et al.* (T2K Collaboration) 2011 *Nucl.Instrum.Meth.* **A659** 106–135 (*Preprint* 1106.1238)
[3] Aguilar-Arevalo A *et al.* (MiniBooNE Collaboration) 2009 *Nucl.Instrum.Meth.* **A599** 28–46 (*Preprint* 0806.4201)
[4] Aliaga L *et al.* (MINERvA Collaboration) 2014 *Nucl.Instrum.Meth.* **A743** 130–159 (*Preprint* 1305.5199)
[5] Aguilar-Arevalo A *et al.* (SciBooNE Collaboration) 2006 (*Preprint* hep-ex/0601022)
[6] Andreopoulos C *et al.* 2010 *Nucl. Instrum. Meth.* **A614** 87–104 (*Preprint* 0905.2517)
[7] Hayato Y 2009 *Acta Phys.Polon.* **B40** 2477–2489
[8] Llewellyn-Smith C H 1972 *Phys. Rep.* **3**(5) 261–379
[9] Bodek A, Avvakumov S, Bradford R and Budd H S 2008 *Eur.Phys.J.* **C53** 349–354 (*Preprint* 0708.1946)
[10] Altegoer J *et al.* (NOMAD) 1998 *Nucl. Instrum. Meth.* **A404** 96–128
[11] Aguilar-Arevalo A *et al.* (MiniBooNE Collaboration) 2010 *Phys.Rev.* **D81** 092005 (*Preprint* 1002.2680)
[12] Lyubushkin V *et al.* (NOMAD Collaboration) 2009 *Eur.Phys.J.* **C63** 355–381 (*Preprint* 0812.4543)
[13] Nakayama S *et al.* (K2K Collaboration) 2005 *Phys.Lett.* **B619** 255–262 (*Preprint* hep-ex/0408134)
[14] Aguilar-Arevalo A A *et al.* (MiniBooNE Collaboration) 2010 *Phys.Rev.* **D81** 013005 (*Preprint* 0911.2063)
[15] Aguilar-Arevalo A *et al.* (MiniBooNE Collaboration) 2011 *Phys.Rev.* **D83** 052007 (*Preprint* 1011.3572)
[16] Aguilar-Arevalo A *et al.* (MiniBooNE Collaboration) 2011 *Phys.Rev.* **D83** 052009 (*Preprint* 1010.3264)
[17] Nieves J, Simo I R and Vacas M J V 2011 *Phys. Rev. C* **83**(4) 045501
[18] Martini M, Ericson M, Chanfray G and Marteau J 2009 *Phys.Rev.* **C80** 065501 (*Preprint* 0910.2622)
[19] Graczyk K M and Sobczyk J T 2008 *Phys.Rev.* **D77** 053001 (*Preprint* 0707.3561)
[20] Morfin J G 2013 *J.Phys.Conf.Ser.* **408** 012054

ORCA: measuring the neutrino mass hierarchy with atmospheric neutrinos in the Mediterranean

Véronique Van Elewyck, for the KM3NeT Collaboration

Laboratoire Astroparticule et Cosmologie (APC), Université Paris Diderot, CNRS/IN2P3, CEA/Irfu, Obs. de Paris, Sorbonne Paris Cité, France

E-mail: elewyck@apc.univ-paris7.fr

Abstract. Since the measurement of the mixing angle θ_{13}, the determination of the neutrino mass hierarchy has become a central challenge of neutrino physics. Recent studies have pointed out that it could reveal itself in the atmospheric neutrino sector, where oscillations are affected by Earth matter effects. This contribution reports on the ORCA feasibility study for such a measurement with an underwater Cherenkov detector based on the technology developed for the KM3NeT neutrino telescope. The baseline performances are discussed for a reference detector with 50 instrumented lines. Preliminary projections, based on the muon channel only, indicate that a $3 - 5\,\sigma$ significance measurement is within reach of a detector with an exposure of the order of 20 Mton years. Further improvement is expected to come from the electron channel, which is currently under study.

1. Introduction and motivations

In the past decade, a large variety of experiments have provided compelling evidence for neutrino oscillations, implying the existence of non-zero neutrino masses. The values of all mixing angles and squared-mass differences in the standard, 3ν, oscillation scheme are by now extracted from global fits of available data with a reasonable precision [1, 2, 3]. The recent measurements of a relatively large value for the mixing angle θ_{13}, which drives the $\nu_\mu \leftrightarrow \nu_e$ transition amplitude, is an asset for the subsequent searches for the remaining unknowns in the neutrino sector, and in particular for the determination of the neutrino mass hierarchy (NMH, normal: $m_1 < m_2 < m_3$ or inverted: $m_3 < m_1 < m_2$). This question is of fundamental importance to constrain the theoretical models that seek to explain the origin of mass; it also impacts the performances of next-generation experiments looking for a leptonic CP-violating phase δ, the absolute value of the neutrino masses, and their Dirac or Majorana nature.

A high-significance ($\geq 5\,\sigma$) NMH determination remains challenging and could take as long as 15 to 20 years even within optimistic estimates [4, 5, 6]. The reach of neutrino beam experiments depends on the still unknown value of δ_{CP}, while atmospheric neutrino experiments suffer from the larger systematics uncertainties and relatively low statistics. The latter probe $\sim 1 - 100$ GeV $\nu_e \leftrightarrow \nu_\mu$ oscillations over a wide range of baselines and neutrino energies, accessing thereby the richness of matter effects that arise in the propagation of neutrinos through the Earth and that reflect in the energy vs. zenith angle oscillogram patterns (see e.g. collective references in [5]). In this context, megaton-scale underwater/ice detectors similar (but denser) to the existing neutrino telescopes and with an energy threshold around the GeV have been proposed as possible contenders for the NMH measurement [7, 8].

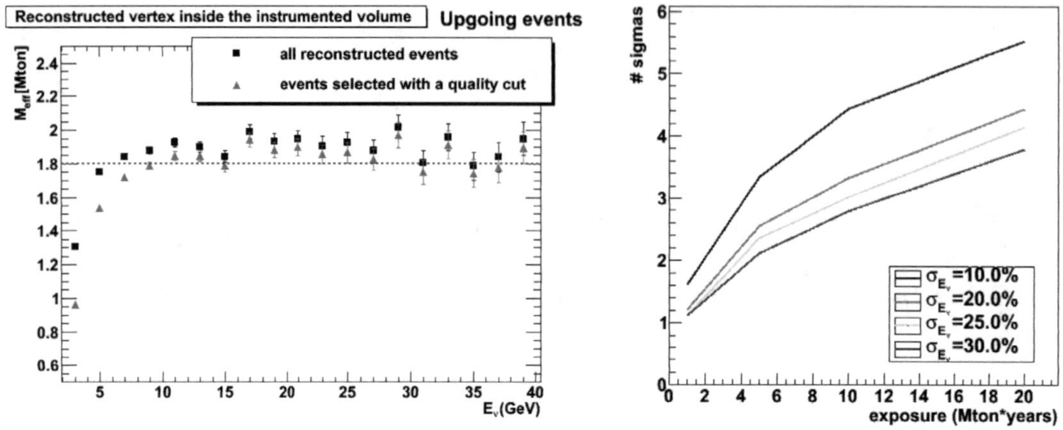

Figure 1. Left: Effective volume (in Mtons of sea water) of the ORCA reference detector as a function of the neutrino energy E_ν, for all reconstructed and for the selected events. The dashed blue line indicates the instrumented volume. **Right:** Significance of the NMH measurement as a function of exposure for different (Gaussian) hypotheses on the neutrino energy resolution.

2. The ORCA feasibility study

ORCA *(Oscillation Research with Cosmics in the Abyss)* is a proposal to conduct a deep-sea water Cherenkov experiment dedicated to the NMH measurement with atmospheric neutrinos in the $1-100$ GeV range. The detector will be a dense array of multi-PMT digital optical modules (DOMs) [9] using technology developed for KM3NeT, the future multi-km^3 neutrino telescope in the Mediterranean Sea [10]. The DOMs are distributed along flexible strings anchored to the sea bottom and held up by a buoy; each of them is equipped with 31 3-inch photomultipliers. The final design of the detector may still depend on the outcome of a dedicated optimisation study which is currently being performed using an overdense detector with maskable DOMs.

The studies presented here have been conducted with a reference detector based on a realistic design in view of the technical and deployment constraints; its cost is estimated at $\sim$ 20 M€. The total height and horizontal spacing of the lines are indeed limited by the deformation of the lines induced by sea currents, and the accuracy of the deployment on the sea bed is, from the ANTARES experience, about 5m [11]. The reference detector comprises 1000 optical modules (OM) distributed on 50 lines arranged in a semi-random pattern inside a circular footprint, with a typical interline spacing of 20m and an inter-OM spacing of 6m. It corresponds to an instrumented water mass of 1.8 Mton. Detailed Monte-Carlo simulations are being conducted to address the main performances of the detector: the trigger and event selection efficiencies, the energy and angular resolutions, the flavour identification capabilities (in particular ν_μ vs. ν_e), the strategies for background rejection and the impact of systematics effects [12].

Studies have so far concentrated on the muon channel, $\nu_\mu N \to \mu X$, with up-going neutrinos (*i.e.* zenith angle $\theta > 90°$). The event reconstruction currently provides the muon track length and direction, the vertex position and a quality parameter Λ. Well-reconstructed events are selected by cutting on Λ and requiring the reconstructed vertex to be inside the instrumented volume. The corresponding effective volume is shown in Fig. 1 (left panel) as a function of the simulated neutrino energy E_ν.A median angular accuracy better than 15° (resp. 10°) above 1 GeV (resp. 5 GeV) has been achieved for the muon neutrino; such a resolution is already mostly dominated by the intrinsic ν_μ to μ angle, demonstrating the capabilities of water as a particle tracker medium. The interaction vertex can be reconstructed with an accuracy of a few meters, which is also an asset for muon background rejection. The energy estimate is currently based on the reconstructed muon tracklength of (semi-)contained events; in the region below 10 GeV where most muon tracks are contained, the typical energy resolution is around 30% $\times E_\nu$. It is anticipated that this resolution will improve once the information from the associated cascade will be included in the reconstruction algorithm.

A likelihood ratio hypothesis test based on pseudo-experiments with a concomitant fit of the oscillation parameters is employed to compute the sensitivity of the experiment for the NMH measurement. Results are shown in Fig 1 (right panel) for different assumptions on the neutrino energy resolution; they can be taken as realistic but on the optimistic side, as the impact of background from atmospheric muons and from other neutrino reactions were neglected so far. A future improvement in the muon channel might still be possible by using the inelasticity parameter y for $\nu/\bar{\nu}$ separation, as suggested in [13].

It has recently become evident that including the electron channel, $\nu_e N \rightarrow eX$, in the analysis can substantially contribute to the significance of the NMH measurement. Although the statistics for ν_e-induced events is smaller, the difference between hierarchies in this channel and indeed largely energy-independant in the energy range of interest, and therefore more robust against detector-related systematics. A thorough study of this channel requires good $\nu_e - \nu_\mu$ discrimination capabilities, and efficient reconstruction algorithms for cascade-like events. Preliminary studies on the cascade channel (from all neutrino flavors), including resolution effects on energy and zenith angle, contributions from neutral current interaction channels ad ν_τ events, indicate that this channel could provide a substantial improvement in sensitivity, as it is the case for the PINGU in-ice proposal [14]. The overall performance of ORCA is still under study.

3. Conclusions and outlook

ORCA has investigated the feasibility of determining the NMH in a densely instrumented water Cherenkov experiment using the technology of KM3NeT, by measuring the energy and zenith distribution of atmospheric neutrinos. Preliminary studies based on the muon channel alone have shown that a $3 - 5\,\sigma$ significance measurement could be within reach of a detector with an overall exposure of 20 Mton year. Including the cascade channel sensitive to ν_e is likely to further improve the prospects, turning ORCA into a competitive experiment in terms of cost and timescale. In addition, the possibility of a neutrino beam from Protvino to the ORCA site [15] should also be investigated further as it would be close to the magical baseline where the separation between normal and inverted hierarchies is optimal.

References

[1] Beringer J *et al.* [Particle Data Group] 2012 Review of Particle Physics, section 13: Neutrino mass, mixing, and oscillations, *Phys.Rev.* D **86** 010001
[2] de Gouvea A *et al.* [Intensity Frontier Neutrino Working Group Collaboration] 2013 *Preprint* arXiv:1310.4340
[3] Capozzi F *et al.* 2013 Status of 3ν oscillation parameters, circa 2013 *Preprint* arXiv:1312.2878
[4] Barger V*et al.* 2012 *Phys. Rev. Lett.* **109** 091801
[5] Ghosh A, Thakore T and Choubey S 2013 JHEP **1304** 009
[6] Blennow M, Coloma P, Huber P and Schwetz T 2014 *JHEP* **1403** 028
[7] Mena O, Mocioiu I and Razzaque S 2008 *Phys. Rev.* D **78** 093003
[8] Akhmedov E K, Razzaque S and Smirnov A Y 2013 *JHEP* **1302** 082 [Erratum-ibid. **1307** 026].
[9] Adrián-Martínez S *et al* [KM3NeT Collaboration] 2014 Deep sea tests of a prototype of the KM3NeT digital optical module *Preprint* arXiv:1405.0839
[10] Margiotta A for the KM3NeT Collaboration 2014 *JINST* **9** (2014) C04020; see also http://www.km3net.org
[11] Ageron M *et al* [ANTARES Collaboration] 2011 *Nucl. Instrum. Meth.* A **656** 11
[12] Katz U F for the KM3NeT Collaboration 2014 The ORCA Option for KM3NeT *Preprint* arXiv:1402.1022
[13] Ribordy M and Smirnov A Y 2013 *Phys. Rev.* D **87** 11 113007
[14] Aartsen MG *et al* [IceCube-PINGU Collaboration] 2014 Letter of Intent: The Precision IceCube Next Generation Upgrade (PINGU) *Preprint* arXiv:1401.2046 [physics.ins-det].
[15] Brunner J 2013 Counting Electrons to Probe the Neutrino Mass Hierarchy *Preprint* arXiv:1304.6230

Topical Research Meeting: Prospects in Neutrino Physics (NuPhys2013) IOP Publishing
Journal of Physics: Conference Series **598** (2015) 012034 doi:10.1088/1742-6596/598/1/012034

The SuperNEMO neutrinoless double beta decay experiment

Cristovao Vilela (on behalf of the NEMO collaboration)
Department of Physics and Astronomy, University College London, Gower Street, London WC1E 6BT
E-mail: cvilela@hep.ucl.ac.uk

Abstract. SuperNEMO is an experiment currently in the construction phase with the aim of searching for neutrinoless double beta decay, a beyond the standard model lepton number violating process. It employs the same strategy as its predecessor NEMO-3, with a tracker and calorimeter surrounding a thin foil of source isotope. Twenty modules are planned, yielding 100 kg of ^{82}Se as source, with a sensitivity to a half-life of 10^{26} years. The first module is intended to demonstrate the very stringent radiopurity requirements, with no background counts being expected in the region of interest of the decay for 2.5 years of data taking. It will hold 7 kg of ^{82}Se, giving it a sensitivity to a half-life of 6.5 x 10^{24} years. The tracking detector is essential for background discrimination and its construction and commissioning are being undertaken in the U.K., primarily at UCL, MSSL (UCL) and The University of Manchester.

1. Introduction

Neutrinoless double beta decay ($0\nu\beta\beta$) is a second order weak process where two nucleus-bound neutrons decay into two protons and two electrons. This process can only occur if neutrinos are Majorana particles and have a finite mass [1].

Majorana neutrinos play a crucial role in see-saw mechanisms which offer an explanation for the smallness of neutrino masses. Lepton number violating processes such as the one being discussed help to elucidate the matter/antimatter asymmetry in the universe [2].

In the simplest case the decay proceeds through the exchange of a light neutrino. Given the chiral nature of the weak interaction this requires a spin-flip of the neutrino, which gives the decay rate a dependence on the mass of the neutrino.

Given the absence of neutrinos in the final state, the distribution of the sum of the electrons' energies is a peak at the Q-value for the decay. An irreducible background for this process is double beta decay with the emission of two neutrinos ($2\nu\beta\beta$), which is allowed by the Standard Model and produces a tail up to the region of interest around the Q-value.

2. SuperNEMO

SuperNEMO is an experiment currently under construction. It employs the strategy proven by the series of NEMO experiments (most notably NEMO-3 [3]) of having a tracker, calorimeter and source as separate entities. NEMO-3 produced the best measurements of $2\nu\beta\beta$ for 7 double beta decay isotopes.

The tracker and calorimeter approach allows for particle identification and measurement of their kinematics. This is used to greatly reduce backgrounds by selecting signal-like event topologies and to generate a number of control channels with alternative event topologies which give accurate measurements of the activities of background sources. This technique also gives SuperNEMO its unique capability to search for alternative $0\nu\beta\beta$ decay mechanisms (*e.g.*, right-handed currents and Majoron emission).

The experiment will consist of several modules. Through an increase in the source mass, an extensive R&D programme and stringent background requirements it is planned to have a sensitivity to $0\nu\beta\beta$ two orders of magnitude better than NEMO-3 (Table 1).

	SuperNEMO	Demonstrator
Source	^{82}Se (^{150}Nd, ^{48}Ca)	^{82}Se
Source mass	100 kg	7 kg
Energy resolution	4 % @ 3 MeV (FWHM)	
^{222}Rn activity in tracker volume	< 0.15 mBq/m^3	
Sensitivity to $0\nu\beta\beta$ (90 % CL)	$T^{1/2} > 10^{26}$ years	$T^{1/2} > 6.6 \times 10^{24}$ years
	$m_{\beta\beta} < 40 - 100$ meV	$m_{\beta\beta} < 200 - 400$ meV

Table 1. Experimental parameters and $0\nu\beta\beta$ sensitivity for the SuperNEMO experiment and its Demonstrator module.

3. The Demonstrator Module

The first SuperNEMO module (the Demonstrator, Figure 1) is now in an advanced stage of construction. Its main objective is to demonstrate the feasibility of achieving the very low background requirements imposed on SuperNEMO. As such it is expected to be the experiment with the lowest background count in the $0\nu\beta\beta$ region of interest for 2.5 years of running.

The detector will take NEMO-3's place in the Modane Underground Laboratory (LSM) and underground commissioning is due to start in late 2014.

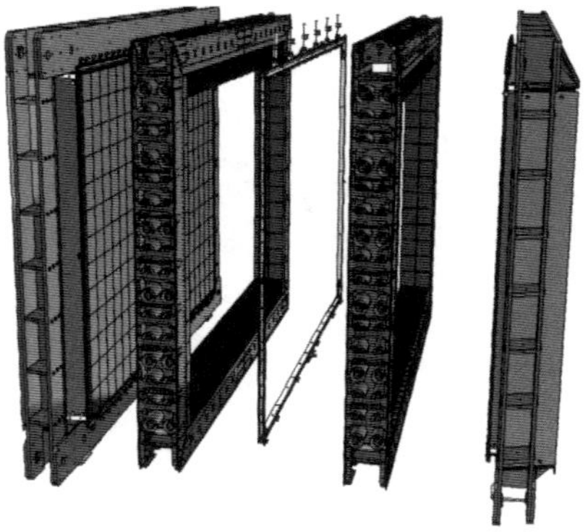

Figure 1. Exploded view drawing of the Demonstrator module. From left to right: calorimeter wall; tracker section; source foil frame; tracker section; calorimeter wall.

Journal of Physics: Conference Series **598** (2015) 012034 doi:10.1088/1742-6596/598/1/012034

4. Source foil

The double beta decay source for the Demonstrator will consist of thin strips of enriched ^{82}Se. A mechanical support material is needed for depositing the source isotope. It is important that the mass of this material is minimised to reduce the background level and minimise energy losses of electrons as they traverse the source. Backing materials such as perforated films and meshes in various materials are currently under study.

A system was designed and is currently under testing which allows for calibration sources to be deployed in the source plane between strips, with minimal additional material inside the detector volume.

5. Calorimeter

The main calorimeter for the first SuperNEMO module consists of 520 plastic scintillator blocks wrapped in PTFE and aluminised mylar, directly coupled to 8" Hamamatsu photomultiplier tubes. These are assembled into eight-block "bricks" which provide mechanical support, magnetic shielding and gas seal.

An extensive R&D programme led to an optimised scintillator block geometry, scintillator material and photomultiplier selection. This resulted in an energy resolution two times better than that of NEMO-3.

6. Tracker

The tracker for the Demonstrator consists of an array of 2034 cells operating in Geiger mode. Each cell contains a central anode wire which is set to $\sim$1.5 kV and is surrounded by 12 grounded field wires in an approximately cylindrical structure 3 m long and 4.4 cm in diameter. Copper rings are placed at both ends of each cell on which a pulse is induced by the arrival of the avalanche. The time difference between an external trigger (from the calorimeter) and a pulse on the anode wire gives the hit's radial distance with an error of 0.7 mm, and the time difference between an anode pulse and pulses on the copper rings gives the longitudinal position of the hit with 1 cm resolution.

The tracker is being assembled in quarters, each quarter containing 504 cells. Each frame is populated with calorimeter modules which enclose the tracking volume. These modules are built with recommissioned NEMO-3 photomultipliers. Tracker quarters will be commissioned at surface at the Mullard Space Science Laboratory, before being shipped to LSM for integration with the rest of the detector and underground commissioning.

7. Conclusion

The SuperNEMO experiment is now at an advanced stage of construction. The first quarter of the tracker has been assembled and populated with calorimeter blocks. Tracker cells are in full production and their insertion in the tracker frame is imminent. The emanation of ^{222}Rn into the tracker volume is being measured at each stage of construction. The calorimeter modules for the main calorimeter are in production. The collaboration is in possession of most of the source required for the Demonstrator, with 5.5 kg of ^{82}Se enriched.

The UK has an important role in this phase of the experiment, having undertaken the construction and commissioning of the tracker.

References

[1] Rodejohann W 2012 *J.Phys.* **G39** 124008 (*Preprint* 1206.2560)
[2] Fukugita M and Yanagida T 1986 *Phys.Lett.* **B174** 45
[3] Arnold R *et al.* 2005 *Nucl. Inst. Meth.* **A536** 79

Using MiniBooNE NCEL and CCQE cross section results to constrain 3+1 sterile neutrino models

C Wilkinson, S Cartwright and L Thompson

Department of Physics and Astronomy, University of Sheffield,
Hicks Building, Hounsfield Road, Sheffield, S3 7RH, United Kingdom

E-mail: `callum.wilkinson@sheffield.ac.uk`

Abstract. The MiniBooNE NCEL and CCQE cross-section measurements (neutrino running) are used to set limits in the $\Delta m^2 - \sin^2 \vartheta_{\mu s}$ plane for a 3+1 sterile neutrino model with a mass splitting $0.1 \leq \Delta m^2 \leq 10.0$ eV^2. GENIE is used, with a relativistic Fermi gas model, to relate E_ν and the reconstructed quantities measured. The issue of uncertainty in the underlying cross-section model and its effect on the sterile neutrino limits is explored, and robust sterile neutrino limits are produced by fitting the sterile parameters and the axial-mass cross-section parameter simultaneously.

1. Introduction

The large axial-mass (M_A) measured by MiniBooNE and other experiments has shown that simple RFG models are inadequate to describe experimental data from quasi-elastic neutrino scattering off nuclear targets. Although there has been a great deal of recent theoretical work developing more sophisticated cross-section models, a clear picture has yet to emerge (a recent summary can be found in [1]). Neutrino oscillation experiments use the measured event rate to infer detailed information about the flux, so a flawed cross-section model may bias results. There have been a number of studies investigating this bias in the context of three-neutrino mixing measurements (see for example [2, 3, 4]). Similarly, such a bias should be investigated for sterile neutrino results, which may be more susceptible as there is generally no way to measure the unoscillated flux.

This work investigates the effect that uncertainty in an RFG cross-section model, with M_A as the only free parameter, has on sterile limits produced by a simple analysis of MiniBooNE NCEL and CCQE cross-section data. It extends the work published in [5], which omitted the CCQE data because of the lack of bin correlations. Limits are set in the $\Delta m^2 - \sin^2 \vartheta_{\mu s}$ plane for a 3+1 sterile neutrino model with a mass splitting $0.1 \leq \Delta m^2 \leq 10.0$ eV^2 using a number of different assumptions about the RFG model. We implicitly follow the assertion made in [6], that inflating M_A provides a reasonable description of the data, though it is understood that this inflated M_A value is effectively accounting for additional nuclear effects. We will refer to the inflated axial-mass as $M_A^{\rm eff}$ from now on. A worthwhile extension of this work would be to look at the effect that different cross-section models have on the sterile neutrino limits produced.

There are two choices to be made regarding the simple cross-section model, and we demonstrate that their effects on the sterile limits are significant. The first is whether to sequentially fit M_A^{eff} then the sterile neutrino parameters, which is only statistically sound if M_A^{eff} and the sterile parameters are completely uncorrelated, or fit all parameters simultaneously. The former procedure was used in the MiniBooNE-SciBooNE sterile analyses [7, 8], which used the MiniBooNE measurement of M_A^{eff} as a constrained parameter in the fit, though it was noted that the cross-section and sterile parameters had been found to be nearly uncorrelated. In general, however, it may not be obvious that an experiment which has a prior *in situ* measurement of their cross-section parameters must investigate correlations with the sterile parameters. The second choice is whether to fit a separate M_A^{eff} value for the NCEL and CCQE samples, or whether one value should be used. The correct choice is not clear without a full understanding of the nuclear effects being covered by M_A^{eff}.

It should be noted that constraints from other experiments cannot be used because the nuclear effects being modelled by M_A^{eff} are detector and neutrino energy dependent. Of course, when more sophisticated models emerge which provide a consistent description of all the available experimental data, these issues will be resolved.

2. Analysis method

Relating reconstructed quantities (the published cross-section results) with E_ν requires a cross-section model. Here we make this model dependence explicit by using GENIE to simulate events on the MiniBooNE detector material, CH_2, with our chosen RFG model, and then producing a flux-averaged cross-section prediction to compare with the published results.

A full description of the method used to produce model predictions can be found in [5] for the NCEL dataset. The extension to include CCQE is straightforward, using the same cross-section model with the signal definition given in Table 1. The MiniBooNE flux prediction is given in [13].

Property	**MiniBooNE NCEL**	**MiniBooNE CCQE**
Baseline L (m)	541	541
Average Neutrino Energy (GeV)	0.788	0.788
Energy Range for Measurement (GeV)	$0 \leq E_\nu \leq 10$	$0 \leq E_\nu \leq 3$
Signal Events	$\nu_{\mu,e} + n,p \rightarrow \nu_{\mu,e} + n,p$	$\nu_\mu + n \rightarrow \mu^- + p$
POT	6.46×10^{20}	5.58×10^{20}
Integrated Flux Φ_ν (ν cm^{-2} POT^{-1})	5.22227×10^{-10}	5.16×10^{-10}
Target Material	CH_2	CH_2

Table 1: Summary of the important experimental details for the two samples used in this analysis. Further details describing the NCEL sample can be found in [9, 10], and for the CCQE sample in [11, 12].

3. Fit Details

We follow the fit procedure described in [5], with a modified χ^2 expression, given in Equation 1, which includes the additional CCQE bins.

$$\chi^2(\theta) = \left[\sum_{i=0}^{51}\sum_{j=0}^{51}\left(\nu_i^{DATA} - \nu_i^{MC}(\theta)\right) M_{ij}^{-1}\left(\nu_j^{DATA} - \nu_j^{MC}(\theta)\right) + \left(\frac{\theta_{M_A}}{\sigma_{M_A}}\right)^2\right] \rightarrow \chi^2_{NCEL}(\theta)$$
$$+ \left[\sum_{k=0}^{17}\left(\frac{\nu_k^{DATA} - \xi\nu_k^{MC}(\theta)}{\sigma_k}\right)^2 + \left(\frac{\xi - 1}{\sigma_\xi}\right)^2\right] \rightarrow \chi^2_{CCQE}(\theta), \quad (1)$$

where ξ is the MiniBooNE normalisation factor, varied in all fits, and σ_ξ is the published CCQE normalisation uncertainty of 10.7% [11].

Note that for sequential fits, additional penalty terms are added to the χ^2 from Equation 1 for each M_A^{eff} parameter included (a full description can be found in [5]). The error and central values for these penalty terms are taken from fits where M_A^{eff} is varied, and no sterile mixing is assumed, the results of which are given in Table 2.

Journal of Physics: Conference Series **598** (2015) 012035 doi:10.1088/1742-6596/598/1/012035

4. Results

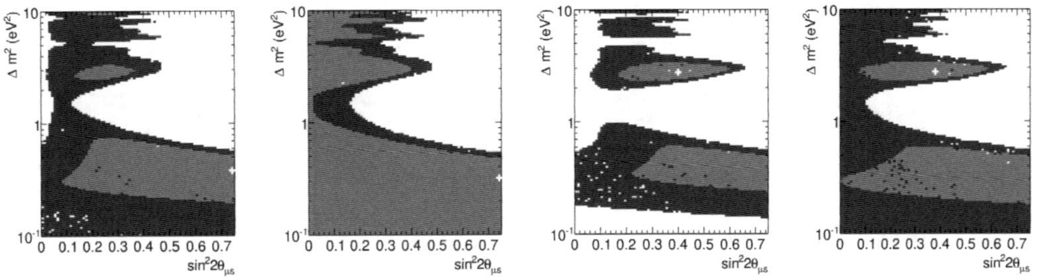

(a) M_A^{BOTH} sequential (b) M_A^{CCQE} & M_A^{NCEL} sequential (c) M_A^{BOTH} simultaneous (d) M_A^{CCQE} & M_A^{NCEL} simultaneous

Figure 1: The exclusion plots produced by joint fits to both NCEL and CCQE datasets. The 90% region is shown in red, the 99% region is shown in blue, and the best fit point is indicated with a yellow cross.

Joint fits to both datasets are shown in Figure 1, sequential and simultaneous fits are shown, both with a common M_A^{BOTH} value for both NCEL and CCQE datasets, and with separate M_A^{NCEL} and M_A^{CCQE} values. The best fit χ^2 and parameter values are given in Table 3.

5. Analysis and Conclusions

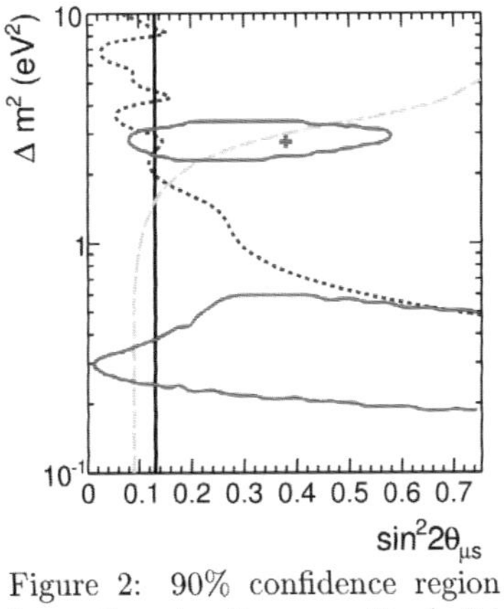

Figure 2: 90% confidence region from the simultaneous fit (solid red line with best fit red cross), with MiniBooNE-SciBooNE ν_μ-disappearance limits [7] (short dashed blue line), atmospheric limits [14] (black solid line), and MINOS NC-disappearance analysis [15] (long dashed green line).

It is clear from Figure 1 that there are differences between sequential and simultaneous fitting, indicating that the sterile and cross-section model parameters are correlated in this case, and therefore the sequential fit is not reliable. We should note that the correlation is strong in the NCEL case, and weak in the CCQE case, so the assertion that they are uncorrelated in [7, 8] is probably reasonable, but in general this cannot be assumed.

There is also a considerable difference between the contours produced when fitting to one or two M_A^{eff} terms, as can be seen by comparing Figures 1c and 1d. The correct choice is not clear, so in Figure 2 we take the more conservative limits, with separate M_A^{CCQE} and M_A^{NCEL} parameters, and compare the 90% confidence regions with existing datasets. It is clear that there is strong disagreement with other limits. Note that in Figure 2 we use the relation $\sin^2 2\theta_{\mu s} \leq \sin^2 2\theta_{\mu\mu}$ to plot other results in the same plane. We interpret the disagreement with other sterile neutrino results as evidence that the cross-section model choice has resulted in tension between the NCEL and CCQE datasets, which was resolved by favouring more sterile mixing. However, it should also be noted that this result is perfectly consistent with the methods used to produce other sterile mixing results, and could also be interpreted as evidence that adding the NCEL dataset gives additional power to constrain the 3+1 model in this plane. The tension with other datasets can be interpreted as evidence that the 3+1 model is insufficient to describe all of the sterile neutrino data available.

	Fit	χ^2/DOF	$\mathbf{M}_A$
This analysis	NCEL	32.1/50	1.24 ± 0.08
	CCQE	20.2/16	1.46 ± 0.05
	Joint	57.0/67	1.40 ± 0.04
MiniBooNE	NCEL [9]	26.9/50	1.39 ± 0.11
	CCQE [11]	—/17	1.35 ± 0.17

Table 2: Best fit values for each of the M_A^{eff} fits performed for this analysis, with published MiniBooNE values for comparison.

Fit Description	Sequential		Simultaneous	
	TWO	ONE	TWO	ONE
χ^2/DOF	47.3/47	46.8/47	44.1/45	44.6/46
Δm^2	0.32	0.38	2.75	2.74
U_{e4}	5.1×10^{-2}	4.8×10^{-7}	3.9×10^{-2}	2.7×10^{-7}
$\text{U}_{\mu4}$	0.50	0.46	0.33	0.34
$\sin^2 2\vartheta_{\mu s}$	0.74	0.74	0.38	0.40
M_A^{NCEL}	1.26	1.38	1.52	1.62
M_A^{CCQE}	1.43	1.38	1.62	1.62
CCQE Norm.	1.10	1.16	1.24	1.26

Table 3: Best fit values for all of the fits performed. Each fit uses either one or two M_A^{eff} parameters. The sequential fits use the relevant M_A^{eff} values and errors calculated in Table 2 in the penalty terms.

[1] Formaggio J A and Zeller G P 2012 *Rev. Mod. Phys.* **84**(3) 1307–1341
[2] Coloma P and Huber P 2013 (*Preprint* 1307.1243)
[3] Mosel U, Lalakulich O and Gallmeister K 2014 *Phys. Rev. Lett.* **112** 151802 (*Preprint* 1311.7288)
[4] Benhar O and Rocco N 2013 (*Preprint* 1310.3869)
[5] Wilkinson C, Cartwright S and Thompson L 2014 *JHEP* **1401** 064 (*Preprint* 1309.1081)
[6] Aguilar-Arevalo A *et al.* (MiniBooNE) 2013 (*Preprint* 1309.7257)
[7] Mahn K *et al.* (SciBooNE, MiniBooNE) 2012 *Phys. Rev.* **D85** 032007 (*Preprint* 1106.5685)
[8] Cheng G *et al.* (MiniBooNE, SciBooNE) 2012 *Phys.Rev.* **D86** 052009 (*Preprint* 1208.0322)
[9] Aguilar-Arevalo A *et al.* (MiniBooNE) 2010 *Phys. Rev.* **D82** 092005 (*Preprint* 1007.4730)
[10] Perevalov D 2009 *Neutrino-nucleus neutral current elastic interactions measurement in MiniBooNE* Ph.D. thesis
[11] Aguilar-Arevalo A *et al.* (MiniBooNE) 2010 *Phys. Rev.* **D81** 092005 (*Preprint* 1002.2680)
[12] Katori T 2008 *A Measurement of the muon neutrino charged current quasielastic interaction and a test of Lorentz violation with the MiniBooNE experiment* Ph.D. thesis
[13] Aguilar-Arevalo A *et al.* (MiniBooNE) 2009 *Phys. Rev.* **D79** 072002 (*Preprint* 0806.1449)
[14] Maltoni M and Schwetz T 2007 *Phys. Rev.* **D76** 093005 (*Preprint* 0705.0107)
[15] Giunti C and Laveder M 2011 *Phys. Rev.* **D84** 093006 (*Preprint* 1109.4033)

AUTHOR INDEX